Daniel Herrmann

Photonics at the Frontiers

Daniel Herrmann

Photonics at the Frontiers

Generation of Few-cycle Light Pulses and Real-time Probing of Charge Transfer in Organic Photovoltaics

Südwestdeutscher Verlag für Hochschulschriften

Impressum/Imprint (nur für Deutschland/only for Germany)
Bibliografische Information der Deutschen Nationalbibliothek: Die Deutsche Nationalbibliothek verzeichnet diese Publikation in der Deutschen Nationalbibliografie; detaillierte bibliografische Daten sind im Internet über http://dnb.d-nb.de abrufbar.
Alle in diesem Buch genannten Marken und Produktnamen unterliegen warenzeichen-, marken- oder patentrechtlichem Schutz bzw. sind Warenzeichen oder eingetragene Warenzeichen der jeweiligen Inhaber. Die Wiedergabe von Marken, Produktnamen, Gebrauchsnamen, Handelsnamen, Warenbezeichnungen u.s.w. in diesem Werk berechtigt auch ohne besondere Kennzeichnung nicht zu der Annahme, dass solche Namen im Sinne der Warenzeichen- und Markenschutzgesetzgebung als frei zu betrachten wären und daher von jedermann benutzt werden dürften.

Coverbild: www.ingimage.com

Verlag: Südwestdeutscher Verlag für Hochschulschriften GmbH & Co. KG
Heinrich-Böcking-Str. 6-8, 66121 Saarbrücken, Deutschland
Telefon +49 681 37 20 271-1, Telefax +49 681 37 20 271-0
Email: info@svh-verlag.de

Approved by: München, LMU, Dissertation, 2011

Herstellung in Deutschland:
Schaltungsdienst Lange o.H.G., Berlin
Books on Demand GmbH, Norderstedt
Reha GmbH, Saarbrücken
Amazon Distribution GmbH, Leipzig
ISBN: 978-3-8381-3070-5

Imprint (only for USA, GB)
Bibliographic information published by the Deutsche Nationalbibliothek: The Deutsche Nationalbibliothek lists this publication in the Deutsche Nationalbibliografie; detailed bibliographic data are available in the Internet at http://dnb.d-nb.de.
Any brand names and product names mentioned in this book are subject to trademark, brand or patent protection and are trademarks or registered trademarks of their respective holders. The use of brand names, product names, common names, trade names, product descriptions etc. even without a particular marking in this works is in no way to be construed to mean that such names may be regarded as unrestricted in respect of trademark and brand protection legislation and could thus be used by anyone.

Cover image: www.ingimage.com

Publisher: Südwestdeutscher Verlag für Hochschulschriften GmbH & Co. KG
Heinrich-Böcking-Str. 6-8, 66121 Saarbrücken, Germany
Phone +49 681 37 20 271-1, Fax +49 681 37 20 271-0
Email: info@svh-verlag.de

Printed in the U.S.A.
Printed in the U.K. by (see last page)
ISBN: 978-3-8381-3070-5

Copyright © 2012 by the author and Südwestdeutscher Verlag für Hochschulschriften GmbH & Co. KG and licensors
All rights reserved. Saarbrücken 2012

Kurzfassung

Die schnellsten bekannten lichtinduzierten Prozesse in der Natur treten auf einer Zeitskala von wenigen Femtosekunden (fs) oder sogar auf einigen hundert Attosekunden (as) auf. Um diese ultraschnellen Licht-Materie-Wechselwirkungen aufzulösen und zu erforschen, sind Lichtpulse von wenigen optischen Zyklen vom extrem Ultravioletten (XUV) bis hin zum Infraroten (IR) erforderlich. Deren Erzeugung stellt schon seit Jahren eine Herausforderung dar und stößt auf breites Interesse für Anwendungen in Physik, Chemie und Medizin.

Im *ersten Teil dieser Dissertation* wird die vielversprechende Methodik der nichtkollinearen optisch parametrischen Verstärkung gestreckter Lichtpulse (NOPCPA) für die Generierung von „few-cycle" Lichtpulsen im Sichtbaren (Vis) und nahen IR (NIR) mit Pulsdauern von 5-8 fs Halbwertsbreite erheblich weiterentwickelt. Grundlegende parametrische Einflüsse, wie die Existenz einer parametrisch induzierten Phase und die Generierung von optisch parametrischer Fluoreszenz (OPF), werden sowohl durch theoretische Analysen und numerische Simulationen, als auch durch konkrete Experimente erforscht. Experimentell werden im Rahmen dieser Arbeit „few-cycle" Lichtpulse mit einer Pulsdauer von 7.9 fs, 130 mJ Energie, bei 805 nm Zentralwellenlänge und einem sehr hohen, „seed"-Puls limitierten Vorpuls-Kontrast von 11 und 8 Größenordnungen bei 30 ps und ca. 3 ps erzielt. Diese stellen derzeit die leistungsstärksten „few-cycle" Lichtpulse weltweit dar und es werden durch diese Arbeit und Kooperationen neue Experimente in der Hochfeld-Physik realisiert.

Zum Einen, ist es mit dem hier beschriebenen Breitbandpulsverstärker gelungen, "quasi-monoenergetische" Elektronen mit Energien mit bis zu 50 MeV zu beschleunigen. Dazu wird der Lichtpuls zu relativistischen Intensitäten von mehreren 10^{19} W/cm^2 in einen Helium-Gasjet fokussiert. Die Elektronen zeigen einen stark reduzierten niederenergetischen Elektronenhintergrund, verglichen mit Beschleunigung durch längere Lichtpulse. Zum Anderen, wurde XUV-Licht bis zur 20. Harmonischen des generierten Lichtpulses aus dem Breitbandpulsverstärker durch dessen „sub-cycle" Wechselwirkung mit Festkörperoberflächen erzeugt. Die Erzeugung von solchen kohärenten hohen Harmonischen verspricht den Bau von kompakteren XUV-Strahlungsquellen, die as-Pulsdauern mit hohen Photonenflüssen XUV-Anrege/XUV-Abfrage Experimente kombinieren würden.

Im Rahmen dieser Arbeit werden darüber hinaus neue, erweiterte Konzepte für noch breitbandigeres NOPCPA über eine Oktave entwickelt und charakterisiert, die die Verwendung von zwei Pumppulsen in einer NOPCPA Stufe und die Verwendung von zwei verschiedenen Pumpwellenlängen in zwei aufeinanderfolgenden NOPCPA Stufen beinhalten.

Im *zweiten Teil dieser Dissertation* werden breitbandige Weißlicht-Spektren und mittels NOPCPA spektral abstimmbare, ultrakurze Lichtpulse verwendet um ein weltweit einzigartiges transientes Absorptionsspektrometer mit Vielkanaldetektion zu realisieren. Dieser neue Anrege-Abfrage Aufbau kombiniert eine sehr breitbandige UV-Vis-NIR Abfrage mit einer hohen Zeitauflösung von 40 fs und hoher Sensitivität für die transiente Änderung der optischen Dichte von weniger als 10^{-4}. Damit ist es in dieser Dissertation zum ersten Mal gelungen den photoinduzierten Ladungstransfer im konjugierten Polymer Polythiophen und in hybriden Polythiophen/Silizium Solarzellen in Echtzeit aufzulösen. Dabei wird eine seit mehreren Dekaden geführte kontroverse Debatte über die Natur der primären Photoanregung in organischen Halbleitern aufgelöst: Exzitonen dissoziieren mit 140 fs Zeitkonstante zu Polaronen (Ladungsträger). Entscheidende Parameter (z.B. strukturelle Ordnung, Ladungsträgermobilität) für die Effizienz der Generierung und Extraktion von freien Ladungsträgern können bestimmt werden, was fundamentales Verständnis für die Optimierung von organischer und hybrider Photovoltaik für zukünftige nachhaltige Energiequellen beisteuert. Weitere Ultrakurzzeit-Experimente an neuartigen organischen Solarzellen sind hier begonnen und aufgezeigt.

List of Publications

This thesis is based on the following publications reprinted in the appendices B1-B10:

1   **Generation of sub-three-cycle, 16 TW light pulses by using noncollinear optical parametric chirped-pulse amplification**
    Daniel Herrmann, Laszlo Veisz, Raphael Tautz, Franz Tavella, Karl Schmid, Vladimir Pervak, and Ferenc Krausz
    Optics Letters 34, 2459-24691 (2009).

2   **Generation of 8 fs, 125 mJ Pulses Through Optical Parametric Chirped Pulse Amplification**
    Daniel Herrmann , Laszlo Veisz, Franz Tavella, Karl Schmid, Raphael Tautz, Alexander Buck, Vladimir Pervak and Ferenc Krausz
    Advanced Solid-State Photonics (ASSP) 2009 paper WA3.

3   **Generation of Three-cycle, 16 TW Light Pulses by use of Optical Parametric Chirped Pulse Amplification**
    Daniel Herrmann, Laszlo Veisz, Raphael Tautz, Alexander Buck, Franz Tavella, Karl Schmid, Vladimir Pervak, Michael Scharrer, Philip Russell and Ferenc Krausz
    The European Conference on Lasers and Electro-Optics (CLEO/Europe) 2009 paper CG1_3.

4   **Investigation of two-beam-pumped noncollinear optical parametric chirped-pulse amplification for the generation of few-cycle light pulses**
    Daniel Herrmann, Raphael Tautz, Franz Tavella, Ferenc Krausz, and Laszlo Veisz
    Optics Express 18, 4170-4183 (2010).

5   **Approaching the full octave: Noncollinear optical parametric chirped pulse amplification with two-color pumping**
    D. Herrmann, C. Homann, R. Tautz, M. Scharrer, P. St.J. Russell, F. Krausz, L. Veisz, and E. Riedle
    Optics Express 18, 18752-18762 (2010).

6   **Approaching the Full Octave: Noncollinear Optical Parametric Chirped Pulse Amplification with Two-Color Pumping**
    Daniel Herrmann, Christian Homann, Raphael Tautz, Laszlo Veisz, Ferenc Krausz, and Eberhard Riedle
    Advanced Solid-State Photonics (ASSP) 2011 paper JWC1.

7   **Role of Structural Order and Excess Energy on Ultrafast Free Charge Generation in Hybrid Polythiophene/Si Photovoltaics Probed in Real Time by Near-Infrared Broadband Transient Absorption**
    Daniel Herrmann, Sabrina Niesar, Christina Scharsich, Anna Köhler, Martin Stutzmann, and Eberhard Riedle
    Journal of the American Chemical Society 133, 18220 (2011).

    (Further publications by the authors about this topic are in preparation.)

Additionally, results from various collaborations are contained in this thesis:

8   **Few-Cycle Laser-Driven Electron Acceleration**
    K. Schmid, L. Veisz, F. Tavella, S. Benavides, R. Tautz, D. Herrmann, A. Buck, B. Hidding, A. Marcinkevicius, U. Schramm, M. Geissler, J. Meyer-ter-Vehn, D. Habs, and F. Krausz
    Physical Review Letters 102, 124801 (2009).

9   **Density-transition based electron injector for laser driven wakefield accelerators**
    K. Schmid, A. Buck, C. M. S. Sears, J. M. Mikhailova, R. Tautz, D. Herrmann, M. Geissler, F. Krausz, and L. Veisz
    Physical Review Special Topics-Accelerators and Beams 13, 091301 (2010).

10 **Toward single attosecond pulses using harmonic emission from solid-density plasmas**
    P. Heissler, R. Hörlein, M. Stafe, J.M. Mikhailova, Y. Nomura, D. Herrmann, R. Tautz, S.G. Rykovanov, I.B. Földes, K. Varjú, F. Tavella, A. Marcinkevicius, F. Krausz, L. Veisz, G.D. Tsakiris
    Applied Physics B 101, 511–521 (2010).

The author of this work received written permission by LMU München and the scientific journals to include parts of the dissertation and copies of the scientific publications in this book.

Contents

Chapter 1: Introduction and Thesis Outline 1
Chapter 2: Physics of Noncollinear Optical Parametric Chirped Pulse Amplification 7
    2.1) Broadband Phase-matching 7
    2.2) Choice of Crystal Material 11
    2.3) Numerical Simulation of NOPCPA 13
    2.4) Optical Parametric Fluorescence 14
Chapter 3: Light-Wave Synthesizer 20 NOPCPA System 16
    3.1) Octave-spanning Supercontinuum Generation 17
    3.2) NOPCPA Chain 20
    3.3) Few-cycle Pulse Characterization 24
        3.3.1) Pulse Compression 24
        3.3.2) Temporal Pulse Contrast 28
Chapter 4: Investigation of Two-beam Pumped NOPCPA 31
Chapter 5: Approaching the full Octave via Two-color Pumping 33
Chapter 6: Applications of Multi-TW, Few-cycle Light Pulses 38
    6.1) Few-cycle Laser-Driven Electron Acceleration 38
    6.2) High-harmonic Generation from Solid Surfaces 41
Chapter 7: Principles of Transient Absorption Spectroscopy 44
Chapter 8: UV-Vis-NIR TA spectrometer with 40 fs time resolution 46
Chapter 9: Charge Generation and Recombination in Hybrid P3HT/Si Photovoltaics .. 50
    9.1) Solar Cell Operation 50
    9.2) Choice of Materials for Hybrid Photovoltaics 52
    9.3) Probing the Ultrafast Photovoltaic Charge Generation Mechanism in Real-time 57
        9.3.1) Quantitative Analysis 60
    9.4) Role of Polymer Structural Order and Excess Energy 61
Chapter 10: Effect of Morphology in DIP/C_{60} Photovoltaics 63
Chapter 11: Conclusions and Outlook 66
Bibliography 70
Appendix 75

Chapter 1: Introduction and Thesis Outline

The fastest known light-induced processes in nature occur on a few femtosecond (fs) or even sub-fs timescale [1-4]. In order to resolve these fast light-matter interactions, few-cycle light pulses from the extreme ultra-violet (XUV) to the infrared (IR) have been highly desired over the last decades [5-9].

The research field of laser physics has experienced tremendous development since the first realization of laser light in 1960 [10], where a ruby rod was pumped by a flashlamp. Nd^{3+}-doped crystals and glasses, gas mixtures (HeNe, CO_2) and semiconductor materials were discovered as gain medium in the early phase. Today, rod, thin disc and fiber lasers based on Yb/Nd:YAG are often pumped by diodes allowing for "all solid state lasers" [11,12]. Laser pulses have been generated via gain- and Q-switching. This approach has been extended by active mode-locking in dye lasers [13] and in the 1990s by Kerr-lens *mode-locking* in Ti:sapphire [14]. This milestone in laser research made fs laser pulses from oscillators available in the Vis-IR spectral range. Today, such laser light sources provide few-cycle, carrier-envelope phase (CEP) stable, few-nJ energy pulses at several MHz repetition rates [15]. In parallel, a technique termed *chirped-pulse amplification* (CPA) was developed to overcome stagnating pulse peak intensities [16], which had been limited by self-action effects of the intense light pulse in optical materials. Since then, ultrashort light pulses up to PW peak powers have been realized using Ti:Sa or Nd:glass [17-19].

With these known gain materials used in multi-stage optically pumped chirped pulse amplifiers the attainable spectral width is limited by either the intrinsic gain bandwidth or even more severely the spectral gain narrowing in going from nJ broadband seed light to the desired Joule output levels. Today, pulses around 20 fs seem to be the lower limit of this approach employing specially designed spectral modulators or employing a hybrid system with parametric amplifiers [20]. To overcome these limitations, frequency broadening in either gas-filled Hollow-core fibers (HCF, [21]) or in filaments [22] is widely used. However, so far the use is limited to up to a few mJ [23,24].

An attractive alternative for the direct generation of high-energy extremely broadband pulses is *optical parametric amplification* (OPA). With a noncollinear geometry and pumping by a frequency-doubled femtosecond Ti:sapphire laser, sub-20 fs visible pulses have been generated at 82 MHz repetition rate [25] and multi-μJ pulses at kHz rates [26-28]. The concept of *noncollinear OPA* (NOPA) was in the following optimized to the generation of 4 fs light pulses in the Vis [6].

Pumping by femtosecond Ti:sapphire pulses limits the attainable output energy, even though extremely short pulses with peak powers of 0.5 TW have been demonstrated [29]. Therefore the use of ps- or even ns-pump pulses was suggested in combination with a chirped pulse approach and was termed *optical parametirc chirped-pulse amplification* (OPCPA) [30-32]. This approach offers the unique advantages of broad gain bandwidth supporting few-cycle light pulses, spectral tunability, high single-pass gain of up to 10^6, negligible thermal load in the nonlinear optical crystal due to the instantaneaous energy conversion and the prospect of energy-scalability. However, a few challenges have to be overcome such as the generation of

1

a broadband seed, pump-to-signal synchronization, accurate dispersion management through pulse stretching and compression, sensitive phase-matching conditions for pump, signal and idler, background emission through amplified optical parametric fluorescence (AOPF) and high requirements for the spatial and temporal properties of the pump. A substantial contribution to the success of parametric amplification was made by the discovery of new attractive nonlinear optical crystals (e.g. borates like BBO) and the development of pump lasers providing few-ps pulses both allowing for higher gain and broader gain bandwidth. All known and technically available pump lasers with ps pulse duration operate around 1050 nm due to the laser active materials used, i.e. Nd^+ or Yb^+. The use of such pump lasers in combination with *noncollinear OPCPA* (NOPCPA) has been envisioned to yield few-cycle, multi-TW light pulses in the Vis-NIR spectral range [33-35]. To realize such ambitious goals, the polarization density

$$\vec{P} = \varepsilon_0 \vec{\chi}^{(1)} \vec{E} + \varepsilon_0 \vec{\chi}^{(2)} \vec{E}\vec{E} + \varepsilon_0 \vec{\chi}^{(3)} \vec{E}\vec{E}\vec{E} + ... \qquad (1.1)$$

of optical materials has to be fully harnessed, which is the source term in the wave equation for a driven wave. The second-order susceptibility $\chi^{(2)}$ leads to wave mixing such as second-harmonic generation (SHG), sum-frequency generation (SFG), difference-frequency generation (DFG) or OPA and requires an anharmonic electron potential (Fig. 1(b)). The third-order susceptibility $\chi^{(3)}$ leads to (among other effects) a light intensity-dependent index of refraction and thereby leading to self-action effects of a propagating light pulse in all optical materials (Fig. 1(a)). All of the nonlinear optical processes shown in Fig. 1 are used in this thesis work to generate supercontinua in the Vis-NIR spectral range, to amplify a broad spectral range supporting few-cycle pulses and to characterize the generated ultrashort light pulses (chapters 2 and 3). Additionally, even higher-order nonlinearities are employed to generate XUV radiation yielding ultrashort light sources from 40 to 1200 nm (chapter 6).

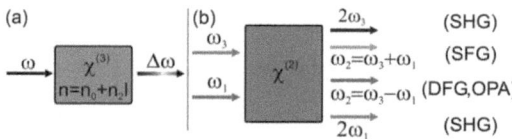

Fig. 1: (a) 3rd-order susceptibility leading to pulse self-action and spectral broadening via the nonlinear index of refraction n_2. (b) 2nd-order susceptibility employed for 2nd-harmonic generation (SHG), sum-frequency generation (SFG) and difference-frequency generation (DFG) / optical parametric amplification (OPA).

In the present thesis work, the generation of sub-three-cycle light pulses at 805 nm central wavelength (7.9 fs intensity FWHM) containing more than 130 mJ of energy at 10 Hz repetition rate is shown by designing a two-stage NOPCPA chain pumped at 532 nm (chapter 3). Nonlinear optical schemes to characterize few-cycle light pulses in the time and frequency domain are developed and the compressibility of the generated 16 TW light pulse very close to the Fourier-limit (7.5 fs) is proven (chapter 3). The compressed pulse shows a high temporal pulse contrast of 10^{-11} at 30 ps, 10^{-8} at 5 ps and 10^{-5} at about 3 ps before the main pulse, respectively, allowing for clean interactions between the main pulse and plasmas. This NOPCPA system is up to date the *most intense few-cycle light source worldwide* [36] and is used to accelerate electrons to 30 MeV and to generate XUV radiation from solid surfaces reaching 40 nm (chapter 6).

Novel schemes to extend the gain bandwidth of NOPCPA towards a full octave in the Vis-NIR spectral range are investigated in this thesis. In a first step, *two-beam pumping* at 532 nm in one BBO crystal is characterized (chapter 4). In a second step, *two-color pumping* in two cascaded NOPCPA stages is realized. It is shown that a nearly octave-spanning bandwidth in the Vis-NIR spectral range can be amplified to mJ-level energies, while the overall bandwidth is composed by parametric amplification in individual NOPCPA stages. The compressibility of such composed spectra due to continously varying spectral phase in the region of overlap is shown. However, at the same time the existence of a *parametric phase* imprinted on the signal during amplification, which strongly depends on the pump wavelength, is proven and certain fundamental consequences of this parametric phase are discussed (chapter 5).

The generation of the shortest light pulses in a broad spectral range from the UV to the NIR allows for probing some of the fastest known processes in nature in real-time. *Primary photoinduced processes* in organic molecules such as energy relaxation, charge or energy transfer and conformational changes occur on timescales from 10 fs to 1 ps. The speed of such fundamental processes is inherently linked to their efficiency. *Ultrafast broadband pump-probe spectroscopy*, employing a tunable NOPA system as pump and broadband NOPA systems [37] or a supercontinuum [38] as probe, is an ideal approach to excite the molecular system on resonance and to probe its optical transitions occuring at various photon energies at the same time (chapters 7 and 8). This approach is chosen in this thesis to reveal the primary photoinduced processes in *hybrid and organic photovoltaics* (chapters 9 and 10).

Thin film photovoltaics of organic semiconductors (e.g. π-conjugated polymers) or inorganic semiconducting nanostructures (e.g. quantum dots, tetra pots, nanocrystals) offer a promising contribution to future sustainable energy supply [39-41]. Such solar cells can be fabricated by low-cost solution processing on flexible substrates via spin-coating or even roll-to-roll assembly. The photoexcitation of a π-conjugated material creates a relatively localized electronically excited state that comes along with a localized relaxation of the molecular structure. This confines the size of a singlet-exciton to only around r=1-3 nm (a few monomer units), while the term *"exciton"* in this context refers to both the electronic excitation and the local bond rearrangements induced by electron-lattice interactions. The Coulomb potential V is to be overcome to fully separate electron and hole at distance r:

$$V = \frac{e^2}{4\pi\varepsilon_r\varepsilon_0 r}, \qquad (1.2)$$

where e is the charge of an electron and ε_0 is the vacuum permittivity. The low initial electron-hole spacing r represents an inherent challenge of organic semiconductors together with the low relative static permittivity ε_r (or dielectric constant) of 2-4 compared to about 12 in case of crystalline silicon. For silicon, V is smaller than the thermal energy 25 meV at room temperature. The exciton binding energy is about 0.1-0.5 eV in case of organic materials, which is much larger than the thermal energy, leading to strongly bound electron-hole pairs. Facing the exciton binding energy, exciton lifetime of $\tau<1$ ns and exciton diffusion length of L~3-14 nm, the need of exciton dissociation within this time and over this distance arises for organic photovoltaics (OPV). This opens the route for combining electron donor and acceptor materials in a thin film heterojunction, where phase segregation between the two components should be on the order of the exciton diffusion length. In such a donor-acceptor combination, electron transfer is referring to the fundamental mechanism in natural photosynthesis [3a]. In photosynthetic reaction centers, after various energy transfer steps, a singlet excited state is localized on a primary donor followed by cascaded downhill electron transfer from the donor LUMO to a neighboring molecular acceptor yielding a long-lived charge-separated state. This charge separation comes with the cost of energy loss as driving force to prevent recombination and thermally activated back-transfer. The same underlying principle prevails for organic and hybrid photovoltaics as approach for *artificial light-havesting*.

Since the establishment of polymers for OPV around 1990 [42,43], some of the highest power conversion efficiencies (around 5%) have been achieved using the polymer *poly(3-hexylthiophene) (P3HT)* in combination with the electron accepting molecule [6,6]-phenyl-C61-butyric acid methyl ester (PCBM) in a thin film bulk heterojunction [44]. P3HT is a semicrystalline polymer that exhibits relatively high charge carrier mobilities which is advantageous for optoelectronic applications [45,46]. Despite the broad absorption range of polymers starting in the UV, the absorbance spectrum of many polymers (such as P3HT and PPV derivatives) is not extending the visible spectral range. This also holds true for PCBM. For this reason, the research of hybrid solar cells has grown rapidly in the recent years, where an organic semiconductor is combined with an inorganic nanostructure offering a broader spectral range of sun light absorption [47-50]. Furthermore, some of the inorganic nanostructures (e.g. CdSe quantum dots) offer the possibility of size-dependent band gap tuning via quantum confinement. *Silicon* is particularly interesting, since it possesses a low band gap of 1.1 eV and therefore absorbs sun light from the UV to the NIR, thus rendering it a very promising alternative to the established PCBM for photovoltaic devices. Furthermore, Si is the second-most available element on earth and (in contrast to Cd) non-toxic.

Efforts to optimize the performance of organic photovoltaic devices find their basis in the fundamental photoconversion processes. The initial photoconversion mechanism of broadband light absorption and the generation of free charge carriers are prerequisites for the operation of an efficient solar cell. Most scientific investigations have focused on the main factors which could determine the efficiency of these photophysical processes (Fig. 2). The *film morphology* has been identified as crucial aspect, so that the device efficiency can be improved significantly by optimizing the processing conditions and film architecture [48,51,52]. While it became clear that the charge carrier mobility is enhanced in semicrystalline polymers [53], studies on the role of *polymer structural order* and therefore *carrier mobility* in the process of charge generation and separation have been emerging only recently [54]. Furthermore, there have been conjectures and indications that *vibrational excess energy* can facilitate dissociation of bound electron-hole pairs (excitons, bound polaron pairs) and the escape of charge carriers from their Coulomb attraction [55-57].

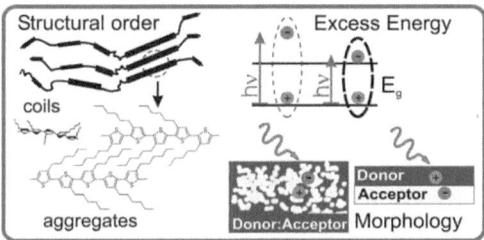

Fig. 2: Potential key factors (polymer structural order, excess energy and sample morphology) determining charge generation efficiency in organic and hybrid photovoltaics.

Despite its central role, the fundamental mechanism of the *initial photoconversion* step of light absorption and charge generation is still highly debated. Nevertheless, there is agreement that in organic blends the process of charge carrier generation (polaron formation) takes place on an ultrafast timescale in the range of 100 fs [1,54,58-61].

Fig. 3 shows the essential two contradicting views of the exact mechanism of free charge generation whose arguments have been exchanged intensely in scientific journals.

One the one hand, some groups (e.g. A. Heeger *et al.*) suggested that light absorption in polymers may directly create mobile electrons and holes by interband π-π^* transitions which would subsequently evolve into Coulombically bound excitons in less than 1 ps [60]. They argue that the *primary photoexcitations* are holes and electrons that are both initially delocalized after interband absorption and undergo charge transfer in a blend with PCBM within ~100 fs (Fig. 3(a)). Similarly, the primary photoexcitation in organic blends has been suggested to be an ultrafast electron transfer on the time scale of 45 fs [1]. Additionally, prompt polaron formation during laser excitation was suggested [59].

On the other hand, some groups (e.g. R. Friend *et al.*) consider the elementary steps involved in the photoconversion process to be light absorption to generate singlet-excitons (Coulombically bound electron-hole pairs) in a donor, the diffusion of the excitation to the internal interface formed by a donor adjacent to an acceptor, and electron transfer from the excited donor to the acceptor eventually forming charge transfer excitons or bound polaron pairs. Ideally, this is followed by the dissociation of the excitons into free charges that escape from the interface (Fig. 3(a)). In this view, the primary photoexcitation upon light absorption is the formation of Frenkel-type singlet-excitons, which dissociate to form polarons on an ultrafast timescale [39,54,58,59,61]. In both opposing models, the free charges in an organic solar cell are a result of an electron transfer from the electron donor to the acceptor and they should preferentially not suffer *geminate or nongemeinate recombination* before being collected at the respective electrodes.

Chapter 1: Introduction and Thesis Outline

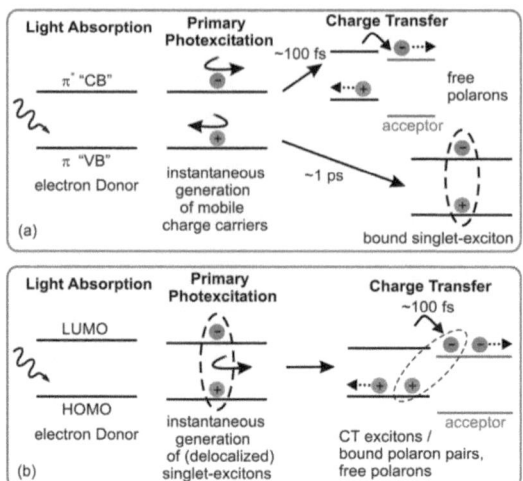

Fig. 3: Debated contradicting views of the primary photoexcitation in organic photovoltaics: (a) instantaneous generation of mobile charge carriers by interband π-π^* transitions on the electron donor and charge transfer within ~100 fs to the acceptor. Subsequent formation of bound singlet-excitons on the ps timescale, and (b) instantaneous formation of delocalized singlet-excitons upon light absorption and subsequent dissociation at a donor-acceptor interface to form free charge carriers within ~100 fs.

The gained understanding within the first part of my thesis work – i.e. generation of tunable ultrashort light pulses as well as supercontinuum generation – enabled the development of a novel ultrabroadband (UV-Vis-NIR) transient absorption spectroscopy setup to probe charge generation and recombination processes in thin *P3HT/Si* photovoltaic films with 40 fs time resolution (chapter 9). This approach allows for monitoring the real-time evolution of excitons and charges as function of excitation intensities, excitation wavelengths and polymer structural order. In P3HT, we observe an instant (<40 fs) formation of singlet-excitons, which subsequently dissociate to form polarons in 140 fs. The retrieved quantum yield of polaron formation through dissociation of delocalized excitons is significantly enhanced by adding Si as an electron acceptor, revealing an ultrafast electron transfer from P3HT to Si. The degree of Coulombic attraction between electron and hole and therefore the separation yield as well as the recombination mechanism depends significantly on polymer structural order.

The developed spectroscopic methods are then also applied to organic photovoltaics consisting of *diindenoperylene (DIP)* combined with the *fullerene C_{60}* (chapter 10). First NIR TA measurements reveal transient signatures of the primary photoinduced species in DIP:C_{60} and also indicate a strong morphology-dependent nanoscopic influence on charge generation and recombination. At the end of this thesis, conclusions and a proposal for future experiments on further promising photovoltaic materials is given (chapter 11).

Chapter 2: Physics of Noncollinear Optical Parametric Chirped Pulse Amplification

Most of the underlying principles of NOPCPA are discussed in detail Ref. [31,62-65] and in the publication in Appendix adding the theoretical discussion of the parametric phase:

Investigation of two-beam-pumped noncollinear optical parametric chirped-pulse amplification for the generation of few-cycle light pulses
Daniel Herrmann, Raphael Tautz, Franz Tavella, Ferenc Krausz, and Laszlo Veisz
Optics Express 18, 4170-4183 (2010).

2.1) Broadband Phase-matching

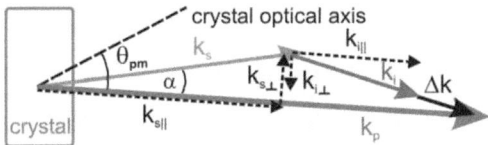

Fig. 4: Wavevector arrangement of pump (p), signal (s), idler (i) and mismatch Δk in case of noncollinear (internal angle α) optical parametric chirped-pulse amplification (NOPCPA) in a uniaxial birefringent crystal, where the index of refraction is adjusted via the crystal optical axis.

A pump photon with angular frequency ω_p propagating in a nonlinear optical crystal, spontaneously or by stimulated parametric emission, decays into two lower energy photons of frequencies ω_s and ω_i. The suffixes p,s and i refer to *pump, signal and idler* waves. The specifc pair of signal and idler frequencies that will result from the parametric process are determined by the energy conservation condition

$$\omega_p = \omega_s + \omega_i, \qquad (2.1)$$

which is intrinsically satisfied. Additionally, the momentum relationship (approximate momentum conservation, Fig. 4) has to be fulfilled:

$$\Delta k = k_p - k_s - k_i \approx 0 \qquad (2.2)$$

with $|k_j| = n(\omega_j, \theta_{pm}, \phi)\dfrac{\omega_j}{c_0} = n(\omega_j, \theta_{pm}, \phi)\dfrac{2\pi}{\lambda_j}$. θ_{pm} is the angle between the propagation vector

and the crystalline optic axis and ϕ is the azimuthal angle. We will see below that this condition can be fulfilled within the limits of *phase-matching*. Due to the presence of *wavevector-mismatch* Δk, an accumulated phase | ΔkL| over the propagation distance L results. For a coherent build-up of the amplified signal the overall accumulated phase slippage should

Chapter 2: Physics of Noncollinear Optical Parametric Chirped Pulse Amplification

be kept low before the signal and idler can convert their energy back to the pump. According to the coupled wave equations, maximum signal amplitude increase is seen for constant phase relationship between the waves with the generalized phase $sin(\Theta) = -1$ [62]:

$$\Theta = \Phi_p - \Phi_s - \Phi_i = -\frac{\pi}{2}. \tag{2.3}$$

Therefore, the phases of the waves have to adjust to maintain high gain through Eq. (2.3). In particular, the phases of the waves are significant, if the pump reaches depletion during signal amplification. At maximum pump depletion Θ reaches zero and changes sign upon reverse of energy flow. For this reason, an additional *parametric phase* is imprinted on the signal wave during amplification to compensate for wavevector-mismatch and to maintain amplification as theoretically outlined in Ref. [31, 66, B4] and experimentally shown in Ref. [B5,B6,66]:

$$\Phi_s(z=L) = \Phi_s(0) - \frac{\Delta k \cdot L}{2} + \arctan\left\{\frac{\Delta k \cdot \tanh(\gamma \cdot L)}{2 \cdot \gamma}\right\}, \tag{2.4}$$

with $\gamma = \sqrt{g^2 - \left(\frac{\Delta k}{2}\right)^2}$, $g \sim d_{eff}\sqrt{\frac{I_p(0)}{n_p n_s n_i \lambda_s \lambda_i}}$ and effective nonlinear optical coefficient d_{eff}, which is half of the 2nd-order susceptibility (Fig. 1). Eq. (2.4) is strictly speaking only valid for the case of low degree of pump depletion through low input signal intensity compared to the input pump intensity. Fig. 6 shows that the parametric signal phase Φ_s plays an important role in optical parametric amplification as it enables broadband amplification. Moreover, it can yield pump intensity-dependent variations of the CEP for the signal. The amplified signal spectral boundaries in this thesis are determined by the effective phase-matching bandwidth shown in Fig. 6. The effective phase-mismatch in a nonlinear optical crystal of length L is calculated according to Ref. [B4] as the sum of crystal-dependent wavevector mismatch ΔkL and amplification-dependent parametric phase (Eq. (2.4)). ΔkL leads to a phase-slippage between pump, seed and idler waves interacting in the crystal. The parametric phase is an inherent phase imprinted on the signal in frequency and space domain during amplification so as to compensate for the phase-slippage and therefore maintain high gain even in areas of significant ΔkL. An accumulated effective phase-mismatch of $\pm\pi$ over the crystal length is acceptable for coherent build-up of the amplified signal in the small-signal gain regime. The compensating effect of the parametric phase broadens the acceptance bandwidth. In the experiments described in the course of this thesis work, the amplified bandwidth in the various NOPCPA stages pumped by several wavelengths and multiple beams with individual phase-matching parameters always match exactly the calculated range based on this analysis. Fig. 11 reveals numerical results of this work, where the signal phase shift due to amplification is shown for increasing degree of saturation and pump depletion, which qualitatively match the calculated parametric phase in such amplifiers. We will discuss the parametric phase in more detail later in chapter 2, 3 and 4 in this thesis. Eq. (2.3) is valid at the beginning of parametric amplification with zero initial idler wave. The idler phase then automatically adjusts to

$$\Phi_i(0) = \Phi_p(0) - \Phi_s(0) + \frac{\pi}{2} \tag{2.5}$$

and thus takes away initial phase changes of signal and pump to ensure maximum initial gain. This is employed to generate CEP-stable light pulses via difference-frequency generation (DFG) when pump and signal are derived from a common light source [67].

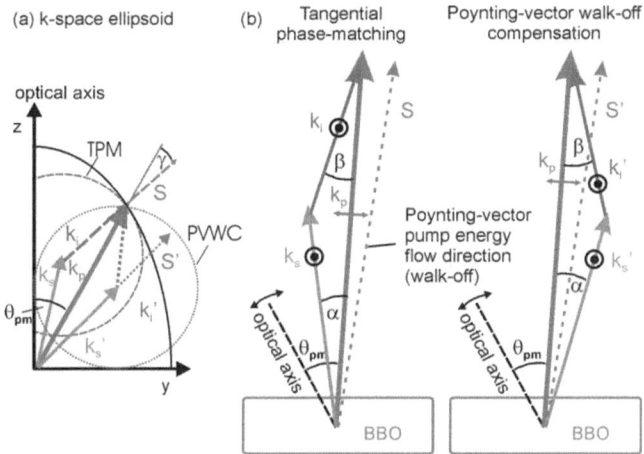

Fig. 5: Two possible phase-matching geometries for type-I NOPCPA in a negative uniaxial crystal ($n_x=n_y=n_o>n_z=n_e$) with nocollinear angle α, phase-matching angle θ_{pm} and walk-off angle γ: tangential phase-matching (TPM) and Poynting-vector (S) walk-off compensation phase-matching (PVWC) with signal (s) and idler (i) polarized in the ordinary uniaxial crystal plane and pump (p) polarized in the extraordinary plane (a) k-space ellipsoid for ideal TPM and PVWC (b) usually nonideal experimental situation with the plane of noncollinear interaction containing the pump polarization.

Fig. 5 shows the phase-matched wavevector alignment for pump, signal and idler waves in both possible phase-matching geometries, namely *Poynting-vector walk-off compensation phase-matching (PVWC)* and *tangential phase-matching (TPM)* [68,69]. The plane of the noncollinear interaction defined by pump and signal wavevector contains the pump polarization and the crystal optical axis. The walk-off is the separation of the extraordinary wave normal from the ray direction due to birefringence [64] and can reduce the parametric interaction length typically to a few mm [68]. Noncollinear geometry can reduce this effect by the signal wave refracting towards the Poynting-vector (PVWC). However, the idler wave-vector then cuts the pump k-space ellipsoid at an acute angle (Fig. 5(a)) leading to a reduced

Chapter 2: Physics of Noncollinear Optical Parametric Chirped Pulse Amplification

angular acceptance compared to TPM, where the idler arc is nearly tangential to the pump ellipsoid. The parallel component of the wave-vector mismatch is insensitive to θ for the ideal case of γ=β. A low angular acceptance is detrimental for highly divergent (focused) beams [69]. Additionally, the PVWC scheme often yields parasitic SHG reducing the OPA efficiency. This is due to the fact that in this case the angle between crystal optical axis and signal beam is close to the optimum phase-matching angle for SHG of the signal. TPM prevents this parasitic by-product but leads to increased spatial pump-signal walk-off also adversely affecting the direction of the amplified signal with respect to the unamplified seed as is discussed in chapter 5 and Ref. [B5]. In both geometries, the participating waves share the same crystal azimuthal angle φ (not critical for phase-matching in uniaxial crystals) which is not the case for the employment of two-beam pumped NOPCPA as discussed in chapter 4 and Ref. [B4].

In general, the wavevector-mismatch can be expanded in a Taylor series:

$$\Delta k(\omega) = \Delta k_0 + \left.\frac{\partial \Delta k}{\partial \omega}\right|_{\omega_0} \Delta\omega + \frac{1}{2}\left.\frac{\partial^2 \Delta k}{\partial \omega^2}\right|_{\omega_0} (\Delta\omega)^2 + O(\Delta\omega)^3, \qquad (2.6)$$

with $\Delta\omega = \omega - \omega_0$. To achieve phase-matching of signal angular frequency ω for a broad bandwidth around the central frequency ω₀, Δk_0 and the first derivatives in Eq. (2.6) should vanish. This can be achieved employing the birefringence of nonlinear optical crystals and introducing an angle α enclosed by pump and signal termed *noncollinear angle*. Applying then the phase-matching condition to Eq. (2.6) according to the detailed derivation in appendix A1 with the assumption of a quasi-monochromatic pump, yields:

$$\upsilon_{gs} = \upsilon_{gi} \cdot \cos(\Omega), \qquad (2.7)$$

with the group velocities $\upsilon_g = \left(\frac{\partial k}{\partial \omega}\right)^{-1}$, Ω=α+β and β is the angle between idler beam and pump beam in case of Δk=0 (Fig. 5). Therefore, matching of the signal group velocity with the idler group velocity in signal direction is necessary to eliminate the first-order derivative of the wavevector-mismatch in Eq. (2.6) and can be achieved in noncollinear optical parametric amplification (NOPA) [63].

To obtain an ultrabroad plateau around the central seed wavelength with small enough wavevector mismatch to allow optical parametric amplification over an even broader signal bandwidth, the second-order wavevector-mismatch in Eq. (2.6) needs to vanish, too. An additional degree of freedom is thus required, such as wavelength-dependent noncollinear angle α(ω) (angular dispersion) or a broadband chirped pump pulse. Alternatively, different approaches are developed and characterized in this thesis work to resolve this challenge and to provide such an ultraboadband gain bandwidth: using two pump beams to amplify adjacent spectral components of the seed bandwidth (*two-beam pumped NOPCPA*) and pumping subsequent amplification stages at different pump wavelengths (*two-color pumped NOPCPA*).

2.2) Choice of Crystal Material

Optical parametric amplification is a nonlinear interaction between three harmonic waves as special case of DFG and can be described by three *coupled wave equations* as outlined in Ref. [31,62,64,65]. Assuming slowly-varying-envelope, zero pump depletion and $\Delta k=0$, an analytic solution for the *small signal intensity gain* can be derived [62,64]:

$$G = \cosh^2(g \cdot L) = \cosh^2\left(d_{eff} \cdot \sqrt{I_p(0)} \cdot L \cdot \text{const.}\right) \tag{2.8}$$

for $\Delta k=0$. There are high requirements that have to be fulfilled by nonlinear optical crystals to be suitable for few-cycle, high-energy OPCPA. On the one hand, the material has to offer a large effective nonlinear optical coefficient (suceptibility) d_{eff}, a high damage threshold and sufficient aperture to provide efficient amplification of high-energy pulses. On the other hand, the crystals have to support the required broad phase-matching bandwidth for few-cycle light pulses at a small noncollinear angle α to minimize spatial walk-off between pump and signal. The material should possess a spectrally broad transparency range from the UV to the IR to prevent two-photon absorption of the pump and to avoid idler absorption. The idler absorption is in general detrimental since it reduces the parametric amplification. However, it was recently shown that idler absorption can be exploited to reduce back conversion in saturated optical parametric amplifiers [70]. Furthermore, the crystal material should show low hygroscopicity and should be coatable with high transmission layers.

Among the availble nonlinear optical crystals, the borate crystals BBO (Beta-Barium Borate), BiBO and LBO (Lithium Triborate) exhibit the best overall characteristics. Especially their high d_{eff} (2.2, 3.2, 0.83 pm/V), large damage threshold (~5-10 GW/cm² for 100 ps pulses at 532 nm) and point of zero GVD (max. of the group velocity) at 1.4-1.6 µm make them unique [71]. As also indicated in Fig. 6, BBO shows the broadest phase-matching conditions in the Vis-NIR from 700 to 1055 nm, but unfortunately its maximum available aperture is currently about 30 mm similar to BiBO. As will be shown in section 3.1, BBO matches best the available supercontinuum seed bandwidth. LBO is available in larger apertures and also shows a broad gain bandwidth (735-1110 nm), which exceeds the NIR content of the seed used in this thesis while covering less of the Vis part. Yttrium Calcium Oxyborate (YCOB) is a promising alternative to BBO as it is available in large apertures and shows a high d_{eff} of 0.98 pm/V and but the gain bandwidth shows reduced NIR content (695-940 nm) similar to BiBO. KDP and DKDP are available in much larger size that makes them suitable for PW-class OPCPA systems, but they suffer from small d_{eff} (~0.24 pm/V), low damage threshold and narrower phase-matching bandwidth compared to borate crystals at the same crystal thickness. For all crystals, the gain bandwidth in the NIR can be extended by adding OPA pumped at 532 nm or NOPA pumping at 1064 nm.

Fig. 6 shows the effective phase-mismatch as sum of wavevector-mismatch ΔkL (blue dotted curves) and parametric phase Φ_s (red dotted curve) yielding the amplified signal bandwdith according to the discussion in section 2.1. BBO (top), LBO (middle) and YCOB (bottom) are considered to indicate their individual potential for high-gain broadband amplification. For this comparison, the crucial factor $d_{eff} \cdot \sqrt{I_p(0)} \cdot L$ in Eq. (2.8) is kept constant to consider the same small signal gain for a parametric preamplifier pumped with a few 10 mJs at 532 nm. Appendix A2 contains formulars for the phase-matching calculations.

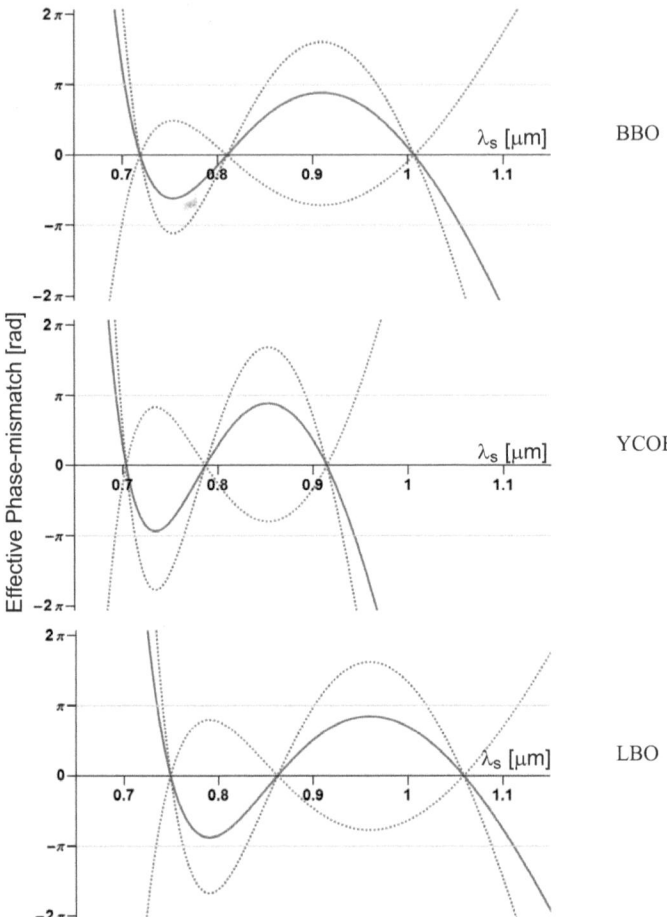

Fig. 6: Effective phase-mismatch (solid purple curves) vs. signal wavelength as sum of ΔkL (blue dotted curves) and parametric phase (red dotted curves). $\pm\pi$ is labeled (solid green lines). Top: L=5 mm uniaxial BBO crystal with α=2.23° (internal), θ_{pm}=23.62°, d_{eff}=2.01 pm/V. Middle: L=10.3 mm biaxial YCOB crystal (XY) taking α=2.77°, ϕ_{pm}=23.32°, d_{eff}=0.98 pm/V, λ_p=532 nm. Bottom: L=12.1 mm biaxial LBO crystal (XY) taking α=1.23°, ϕ_{pm}=13.38°, d_{eff}=0.83 pm/V, λ_p=532 nm for all crystals.

2.3) Numerical Simulation of NOPCPA

This model of negligible pump depletion is useful for a first estimation of NOPCPA parameters and it gives a clear picture for the small gain regime in Eq. (2.8). However, numerical simulations of NOPCPA are required to include pump depletion, *optical parametric fluorescence (OPF)* as well as nonlinear spectral and spatial phase effects. In this way, a more accurate knowledge of the fundamental dynamics in this process can be gained [72-74]. In the present thesis work, numerical split-step simulations of NOPCPA were developed and used to mimic experimental results one the one hand and to study the influence of certain parameters (such as phase-matching conditions, seed energy, pulse durations, spatial beam properties or OPF) on signal amplification. The code was initially described in Ref. [73] and extended to include more wavelength-dependent effects and for pumping with two pump beams [B4].

The numerical code modelling NOPCPA is described in detail in the publication in Appendix:

Investigation of two-beam-pumped noncollinear optical parametric chirped-pulse amplification for the generation of few-cycle light pulses
Daniel Herrmann, Raphael Tautz, Franz Tavella, Ferenc Krausz, and Laszlo Veisz
Optics Express 18, 4170-4183 (2010).

In this approach, the parametric amplification is performed in the time domain (nonlinear step). The coupled wave equations are applied to pump, signal, idler, OPF in signal direction and OPF in idler direction. Moreover, several pump beams yielding several idler beams can be included with their individual phase-matching conditions. The spatial walk-off effect due to the noncollinearity of the type-I parametric process is taken into account by shifting the respective field at each step in the direction defined by the noncollinear angle. Higher-order dispersion and diffraction terms can also be accommodated by adding the corresponding spectral and spatial phase terms in the Fourier domain (linear step). This becomes necessary if the pulse duration is short (dispersion effects) or in the case of small beam sizes (diffraction effects). However, self-phase modulation as well as temporal walk-off due to group-velocity mismatch can usually be neglected in our case because of the long (40 ps) pulses used, and the diffraction effects are not important owing to their weak relative impact for unfocused beams. Parasitic SHG can be neglected by choice of tangential phase-matching in the experiments, where no such losses have been observed. The simulation results of parametric amplification for the experimentally determined input parameters are shown in Fig. 10 for LWS-20 and in Ref. [B4] for two-beam-pumped NOPCPA.

2.4) Optical Parametric Fluorescence

Despite its potential to give a sufficient understanding and description of OPA – as shown in this thesis work – the classical wave theory can not explain the occurance of OPF, i.e. the generation of photons at signal and idler frequencies even in the abscence of seed photons. This quantum phenomenon is a consequence of seeding the parametric amplifier by zero-point mode fluctuations of the vacuum field [75-77]. OPF was initially exploited in context of optical parametric oscillators (OPO) and optical parametric generators (OPG) [68,69]. Its significance as potential source for pulse contrast degradation in OPCPA was considered only recently [B4,73,78-80]: within the pump pulse duration a large incompressible pedestal of amplified OPF (AOPF) can reduce the temporal pulse contrast and contribute to the measured signal and idler energies.

In the quantum mechanical model of parametric amplification proposed by Louisell, Yariv and Siegman [75], signal and idler are two modes of harmonic oscillation of the electromagnetic field that possess a symmetric role in the parametric amplification. The two modes are regarded as coupled by another harmonic oscillator at frequency ω_p equal to the sum of the inherent signal (ω_s) and idler (ω_i) frequencies of the unperturbed oscillators. In this case, the system Hamiltonian is given by:

$$H = H_0 + H_{amp} = \sum_{j=p,s,i} \hbar\omega_j \left(a_j^\dagger a_j + \frac{1}{2}\right) - \hbar\kappa\left[a_s^\dagger a_i^\dagger e^{-i(\omega_p t+\varphi)} + a_s a_i e^{i(\omega_p t+\varphi)}\right], \qquad (2.9)$$

where a_j, $a_j^\dagger$ are the time-dependent photon annihilation and creation operators, κ is the coupling strength (proportional to d_{eff} times the complex amplitude of the pump wave) and φ is the phase of the externally imposed oscillation. The second part of Eq. (2.9) arises from amplification, where an equal number of signal and idler photons are created or annihilated simultaneously at the expense of the annihilation or creation of an equal number of pump photons. The pump field in this case is the incident light on a nonlinear dielectric, which excites the emission of light in signal and idler modes meeting the resonance condition. The equation of motions for a_j, $a_j^\dagger$ (Heisenberg equations of motion) lead to:

$$\frac{da_s}{dt} = -i\omega_s a_s + i\kappa a_i^\dagger e^{-i(\omega_p t+\varphi)} \qquad (2.10)$$

$$\frac{da_i^\dagger}{dt} = i\omega_i a_i^\dagger - i\kappa a_s e^{i(\omega_p t+\varphi)}, \qquad (2.11)$$

which directly yield the Manley-Rowe relations:

$$\frac{d}{dt}a_i^\dagger a_i = \frac{d}{dt}a_s^\dagger a_s, \qquad (2.12)$$

with the photon number operator $N_j = a_j^\dagger a_j$. The integration of Eqs. (2.10), (2.11) gives the solution:

Chapter 2: Physics of Noncollinear Optical Parametric Chirped Pulse Amplification

$$a_s(t) = e^{-i\omega_s t}\left[a_{s0}\cosh(\kappa t) + ie^{-i\varphi}a_{i0}^{\dagger}\sinh(\kappa t)\right], \quad (2.13)$$

$$a_i^{\dagger}(t) = e^{i\omega_i t}\left[a_{i0}^{\dagger}\cosh(\kappa t) - ie^{-i\varphi}a_{s0}\sinh(\kappa t)\right]. \quad (2.14)$$

Now, the average number of signal and idler photons $n_j(t)$ as a function of time is evaluated. At t=0, there are assumed to be n_{s0} signal photons and n_{i0} idler photons as initial state $|n_{s0}, n_{i0}\rangle$ of the system. The evaluation of the expectation value of the photon number operator gives:

$$\langle N_s(t)\rangle = \langle a_s^{\dagger}(t)a_s(t)\rangle$$

$$= \langle n_{s0}, n_{i0}|a_{s0}^{\dagger}a_{s0}\cosh^2(\kappa t) + (1+a_{i0}^{\dagger}a_{i0})\sinh^2(\kappa t) + \frac{i}{2}\sinh(2\kappa t)\left(a_{s0}^{\dagger}a_{i0}^{\dagger}e^{-i\varphi} - a_{s0}a_{i0}e^{+i\varphi}\right)|n_{s0}, n_{i0}\rangle$$

$$= n_{s0}\cosh^2(\kappa t) + (1+n_{i0})\sinh^2(\kappa t)$$

$$(2.15)$$

Moreover, in the vacuum state $|0,0\rangle$ Eq. (2.15) directly yields:

$$\langle N_s(t)\rangle = \sinh^2(\kappa t), \quad (2.16)$$

which is a very interesting result. It states that even in the abscence of initial seed photons, the parametric amplifier spontaneously generates photons at signal and idler frequencies arising from zero-point fluctuations of the respective fields. The rate of this photon generation is equal for signal and idler and follows the same principal dependencies as the small-signal gain in case of phase-matching and in the absence of depletion as stated in Eq. (2.8). This important finding justifies the incorporation of AOPF into numerical simulations of NOPCPA. By summing over all possible frequencies and including phase-matching, the overall output power of OPF can be determined. It turns out to be proportional to d_{eff}, pump intensity and crystal length [77,81], thereby offering a way to determine d_{eff}.

A recently envisioned and more convenient alternative to strict quantum-mechanical treatment of AOPF, are numerical simulations employing stochastic fields. In this way, the implications of some factors such as temporal and spatial walk-off, dispersion, diffraction, crystal length and seed energy on AOPF could be revealed [73,82,83,B4]. Experimentally, the AOPF in parametric amplifiers is often determined by blocking the seed and detecting the signal output. This does not give the actual level of AOPF occurring during parametric amplification, because in this case the OPF and the signal are competing to extract pump energy. However, it gives an upper value to connect the OPF power with the incident pump intensity. Ref. [B4] describes the numerical implementation of AOPF in this thesis work as well as results for using two pump beams. In Fig. 11, more numerical outcome for the case of LWS-20 is added: the role of seed energy on pulse contrast in saturated parametric amplifiers.

Chapter 3: Light-Wave Synthesizer 20 NOPCPA System

The theoretical understanding of NOPCPA outlined in the previous chapter is applied to an actual experimental setup outlined in Fig. 7 with the goal to generate few-cycle light pulses in the NIR containing more than 100 mJ of energy and exhibiting a high temporal pulse contrast sufficient for high-field laser-plasma interactions. This 16 TW parametric amplifier is termed Light-wave synthesizer 20 (LWS-20) and requires development of: (1) a broadband seed source with sufficient energy and temporal pulse contrast to seed moderate gain NOPCPA, (2) adaptive broadband dispersion management with sufficient throughput employing a polarization-insensitive compression, (3) a high-energy relatively compact pump laser providing high-quality pulses at around 525 nm with moderate pulse durations around 100 ps (All technically available high-energy pump lasers with ps pulse duration operate around 1050 nm due to the laser active materials used, e.g. Nd^+ or Yb^+), (4) optical synchronization between seed and pump, (5) proper relay-imaging and spatial filters for an adequate pump beam profile at the parametric amplifiers, (6) careful design of the parametric preamplifier and saturated NOPCPA stage yielding broadband phase-matching supporting few-cycle pulses in the NIR and sufficient gain for the desired final pulse energy, (7) diagnostics to characterize duration, phase and temporal intensity distribution (i.e., shape, contrast) of few-cycle pulses.

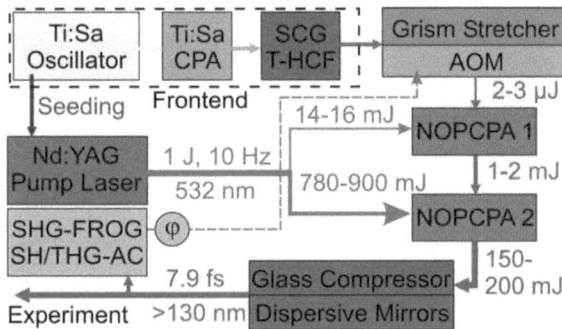

Fig. 7: Schematic layout of the LWS-20 NOPCPA system. Broadband Titan:Sapphire oscillator, CPA: chirped-pulse amplifier, SCG: supercontinuum generation, T-HCF: neon filled tapered hollow-core fiber, grism: prism + grating, AOM: acousto-optic modulator, NOPCPA 1: parametric preamplifier, NOPCPA 2: saturated amplifier (both employing BBO crystals), hybrid compressor, AC: autocorrelators, SHG-FROG: 2nd-harmonic generation frequency resolved optical gating.

The experimental setup and the results of the LWS-20 NOPCPA system drafted in Fig. 7 and discussed in chapter 3 are described in a compact fashion in the corresponding publications in Appendix B1,3:

Generation of sub-three-cycle, 16 TW light pulses by using noncollinear optical parametric chirped-pulse amplification
Daniel Herrmann, Laszlo Veisz, Raphael Tautz, Franz Tavella, Karl Schmid, Vladimir Pervak, and Ferenc Krausz
Optics Letters 34, 2459-24691 (2009).

Generation of Three-cycle, 16 TW Light Pulses by use of Optical Parametric Chirped Pulse Amplification
Daniel Herrmann, Laszlo Veisz, Raphael Tautz, Alexander Buck, Franz Tavella, Karl Schmid, Vladimir Pervak, Michael Scharrer, Philip Russell and Ferenc Krausz
The European Conference on Lasers and Electro-Optics (CLEO/Europe) 2009 paper CG1_3.

Section 3.1 describes the possibilities for broadband seed generation and contains experimental results of *supercontinuum generation* (SCG) in gases. Section 3.2 contains the chosen parameters and the experimental results of the NOPCPA chain. Moreover, it shows results from numerical simulations of the parametric amplifiers and outlines consequences for energy-upscaling, temporal pulse contrast and signal parametric phase. Section 3.3 introduces the reader to the developed few-cycle pulse diagnostics for characterization of the compressed LWS-20 signal pulse in the time and frequency domain. It contains detailed experimental results from 2nd-order and 3rd-order autocorrelation (AC) and SHG frequency resolved optical gating (SHG-FROG). The individual contributions to the temporal pulse contrast landscape are discussed and linked to the numerical simulations.

3.1) Octave-spanning Supercontinuum Generation

In the previous chapter, the high demands of broadband parametric amplification and its practical design rules have been outlined. For the amplification of broadband pulses supporting a few optical cycles, a seed source has to be developed that can provide the required spectral bandwidth. The generation of such a broadband seed is the basis of OPCPA and therefore of high importance. There are a couple of demands for the employed seed generation process. The process of supercontinuum generation (SCG) should yield a stable seed spectral bandwidth with low spectral fluctuations, which is at least as large as the phase-matching bandwidth of the designed NOPCPA stages. The actual SCG process has to keep the spectral phase modulations to a reproducible extend such that compression of the amplified signal pulses close to the Fourier-limit is possible with the currently possible approaches for dispersion management. Further, the SCG scheme needs to provide the required seed energy prior to parametric amplification in a seed beam of high spatial quality. Finally, the experimental setup should not consume an excessive amount of effort to be employed in large-scale NOPCPA systems used for diverse experiments.

There are a couple of SCG (also termed *white-light generation*) mechanisms employed for various CPA and OPCPA/OPA laser systems. First, direct seeding of OPCPA with broadband

Ti:Sa oscillators has been used [34,84], but led to low signal pulse contrast due to its low seed pulse energy (≤nJ) and the ASE originating from the oscillator [98].

Second, spectral broadening via *ionization* and *self-phase modulation* through focusing the output of a CPA laser into bulk material such as YAG or Sapphire [85]. This mechanism offers the prospect of broad bandwidths from the UV to the NIR depending on the crystal and driver wavelength, ease of alignment, high output stability and long lifetime. In general, 1-2 µJ of fundamental light is focused into the crystals currently yielding a few 10 nJ of total output energy at the most [85,86]. Furthermore, the spectral intensity and phase within the SC around the fundamental light is highly modulated and has to be blocked. In fact, this spectral part is not compressible and shows a time jump of about 400 fs between the broadened spectral components below and above the driver. This limits the usage of this SCG scheme to the high-energy or low-energy spectrally broadened part around the fundamental. In case of using a Ti:Sa CPA driver laser, the employable bandwidth would then be 450-750 nm or 800-1600 nm [85]. The high-energy part has successfully been used for Vis few-cycle NOPA pumped at 400 nm [6]. At least the spectral ristrictions can be circumvented by employing a driver laser at above 1 µm (e.g. Yb:YAG) that would provide a nJ seed spectrum from 500 to 1000 nm. The seed energy being rather too low for high-contrast, low-AOPF, high-energy OPCPA [73], this bandwidth matches the phase-matching bandwidth of the Vis-NIR few-cycle NOPCPA scheme used in this thesis work (Fig. 6, [B1-B3]) and that of few-cycle NOPA setups [63,86].

Third, spectral broadening by ionization and self-phase modulation in gases. Here, an energetic (100 µJ up to a few mJ) ultrashort light pulse is tightly focused into a noble gas such as neon, argon or krypton. Compressible octave-spanning output spectra yielding few-cycle pulses containing up to a few mJ energies have been shown to date. There have been two different architecutures. On the one hand, *filamentation* in argon has been investigated. Filamentation is understood as competition between multiphoton ionization that leads to plasma induced (free electron density N_e) beam defocusing

$$\Delta n(r,t) = -\frac{N_e(r,t) \cdot e^2}{2\varepsilon_0 m_e \omega^2}, \tag{3.1}$$

diffraction and the Kerr effect (Fig. 1) that renders self-focusing and self-phase modulation (SPM) [87]. The modulation of the phase $\varphi(r,t) = n(r,t) \cdot (\omega/c) \cdot z$ creates new frequencies via $\omega = -d\varphi/dt$, where in the leading (trailing) edge of the pulse the frequencies are shifted to the red (blue) applying a positive chirp. For a Gaussian pulse, SPM can be approximated by purely even-order phase terms applied to the initial phase and broadens the bandwidth of the pulse if it was initially not negatively chirped. Otherwise, a spectral narrowing would result. During propagation in the nonlinear medium, the pulse intensity maximum is delayed with respect to its edges leading to self-steepening of the trailing edge and an asymmetric spectrum extending more towards the blue. While self-phase modulation and self-steepening determine the spectrally broadened bandwidth of the pulse, the competing spatial effects can induce a self-guiding of the incident beam over more than several Rayleigh lengths. Contradicting reports on filamentation have been made, raising the question of its day-to-day applicability in complex laser systems. Although octave-spanning output spectra could be measured after an argon filament [22, 88], the output beam quality in terms of spatial chirp, maximum achievable energy, spectral fluctuations and compressibility have been under debate. An alternative approach is to extend the length of nonlinear interactions where the intensity remains high by guiding the incident beam in a *hollow-core fiber (HCF)* filled with noble gas [21,23,24,B1-3,B5]. Recently, cascaded combinations of filaments and/or HCFs have been investigated [89]. The HCF scheme allows using a larger incident beam size, noble gases (such as neon) with lower nonlinear index of refraction n_2 and correspondingly a higher incident power. Altogether, the external guiding might offer for the output the prospect of a higher energy, larger bandwidth, reduced spatial transverse deviations and a more stable mode of operation.

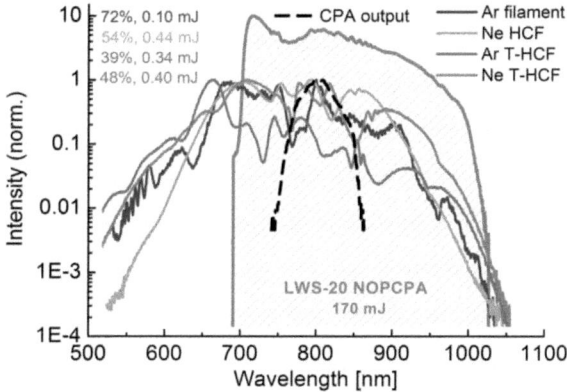

Fig. 8: Octave-spanning supercontinuum generation (SCG) from around 500 to 1050 nm (at 30 dB) in conventional (HCF) or tapered (T-HCF) hollow-core fibers as well as gas filaments filled with neon or argon gas. For each scheme, the relative throughput and the output energy is given. The generated broadband seed before the stretcher is compared to the incident CPA pulse spectrum and the broadest amplified signal spectrum (695-1020 nm, without the unamplified seed components) after the 2nd NOPCPA stage of LWS-20 (5 mm type-I BBO pumped at 532 nm) indicating the parametric gain bandwidth.

For LWS-20 we compared filamentation in argon gas at around 1 bar absolute pressure, broadening in conventional 1 m long HCF with 250 μm inner diameter filled with neon gas (2 bar) and 1.1 m long tapered HCF (T-HCF) with 210 μm smallest inner diameter filled with argon (<1 bar) or neon (2 bar). The smaller inner diameter was used to enhance the intensity along the nonlinear interaction length, while the T-HCF had a 10 cm long taper at the entrance to enhance the coupling of the incident focused beam. The incident pulse was a nearly

Gaussian 25 fs (FWHM) pulse at 800 nm containing up to 0.85 mJ of energy. The incident pulse spectrum and the output spectra of the employed SCG schemes are shown in Fig. 8, together with the throughput characteristics for the case of an adequate output beam. The best performance for using HCFs was achieved using neon gas at around 2 bar. The highest throughput (54%) was achieved for the conventional HCF with narrower output spectrum (530-1000 nm at 30 dB) compared to using the T-HCF, where the spectral broadening could exceed an octave (<500-1050 nm) with the cost of slightly reduced throughput (48%). The argon filament showed a decent spectral broadening and a high throughput but, despite loose focusing, could only be operated in a stable manner at relatively low incident energy. Further investigations can be the subject of future research. The T-HCF filled with neon gas was overall determined to be the best choice as seed source for the LWS-20 NOPCPA system, as the generated bandwidth just matches the phase-matching bandwidth of the designed NOPCPA stages around 1050 nm while it exceeds it in the Vis as shown in Fig. 6 [B1,B2].

The employed dispersion management for LWS-20 was based on a combination of a grism stretcher [35], AOM (Dazzler), glass blocks and dispersive mirrors. The spectral transmission limited the stretched seed bandwidth to 650-1000 nm containing 3 µJ of energy [B1-B3]. That energy is significantly more than compared to seeding with the stretched output of a broadband oscillator or a SC generated from bulk material [35,84,85]. The latter was employed to develop a novel Vis-NIR SC from a combination of YAG and CaF_2 for the broadband ultrafast transient spectroscopy of solar cells in the second part of this thesis. As will be shown in the following sections, the broadband Vis-NIR seed bandwidth from the T-HCF will just be enough to fill up the entire gain bandwidth using two-color pumped NOPCPA [B5] for the generation of sub-5 fs pulses.

3.2) NOPCPA Chain

In the preceeding section we have shown how the necessary seed bandwidth and energy is made available through spectral broadening in hollow-core fibers (Fig. 8). The proper design of the broadband, high-energy parametric amplifiers are outlined in the following. As shown in detail in Ref. [B1-B3], the LWS-20 NOPCPA system also employs a commercial cutting edge (and therefore not turn-key) flashlamp-pumped Nd:YAG pump laser providing 78 ps (FWHM), 1 J pulses at 532 nm after type-II SHG in KDP at 10 Hz repetition rate. This laser is seeded by the self-frequency shift of the broadband Ti:Sa output. The pump laser output beam properties are Gaussian in time and super-Gaussian in space. The output is relay-imaged with vacuum telescopes onto the 5 mm long wedged BBO crystals in the first and second NOPCPA stage, respectively. The seed is stretched with a combination of grism stretcher and AOM [35] leading to 2-3 µJ seed pulses typically ranging from 650-1020 nm (the short wavelength part of the HCF output is cut here) exhibiting a group delay (GD) of about 40 ps between these spectral boundaries. This GD is close to the optimum for parametric amplification with the employed pump pulse as shown in Ref. [B5] in the course of this thesis work and in agreement with numerical simulations [74]. Both NOPCPA stages are operated in tangential type-I phase-matching geometry with the pump polarization contained in the plane of noncollinear interaction, which is defined by the signal and pump beam. This was chosen to circumvent parasitic SHG of the signal. The first amplifier is pumped with 14-16 mJ (>10 GW/cm^2 peak intensity) at 532 nm and led to 1-2 mJ signal energy, which corresponds to around 10% pump to signal energy conversion efficiency (absolute signal energy gain devided by incident pump energy).

Fig. 9 shows that the signal spectrum (blue) ranging from 695-1000 nm and shows a more balanced spectral energy density distribution compared to the unamplified seed (red) but still contains its distinct spectral features. The second amplifier is seeded with the output of the first amplifier and is pumped with 780 mJ (~8 GW/cm^2 peak intensity) at 532 nm. This NOPCPA stage is operated close to saturation and led to up to 170 mJ signal energy corresponding to 22% pump to signal energy conversion efficiency, which is among the most efficient NOPCPA setups. The RMS of the signal energy is about 3%, which is almost limited by the 2% RMS of the pump pulse energy. The measured signal spectrum shown in Fig. 9 (yellow) ranges from 695-1020 nm and shows less pronounced spectral energy density variations compared to the unamplified seed and the signal from the first amplifier. The saturation in the second amplifier actually smoothens the spectral energy density and leads to enhanced NIR content. The high signal gain of the NOPCPA chain therefore allows usage of the exponentially decreasing seed light at the spectral boundaries.

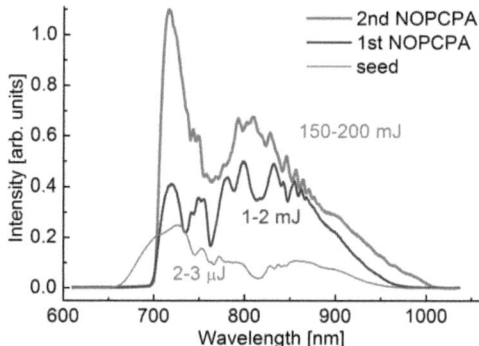

Fig. 9: Spectral energy density (arbitrarily scaled) and total energy of the unamplified stretched seed (red) and the signal after parametric amplification in the 1st (blue) and 2nd (yellow) NOPCPA stage of LWS-20. Pump energies: 14-16 mJ (1st stage), 780-900 mJ (2nd stage).

Later on, the applied pump energy in the second amplifier could be increased to about 900 mJ by choosing custom-made HR mirrors in the pump relay imaging. After carefully choosing the pump beam size and seeding with 2 mJ from the first amplifier, up to 200 mJ signal energy with the same pump to signal conversion efficiency could be obtained. This proofs the scalability of the chosen NOPCPA concept at least in a certain parameter regime. Fig. 10(a) shows the simulated parametric intensity gain spectrum for the case of a partially saturated 2nd NOPCPA stage (2 mJ seed energy, black) and for the case of strong back conversion (5 mJ seed energy, red) at around the signal central wavelength of 800 nm, where Δk=0 was aimed for (compare the phase-matching curve in Fig. 6), also leading to an overall energy reduction (Fig. 10(b)). Apart from the experimentally determined beam properties, the condition for the broadest phase-matching (θ_{pm}=23.62°, α=2.23°) and a proper pump-signal delay were chosen for the numerical simulation.

The simulation in Fig. 10 predicts a signal spectrum starting just below 700 nm and ranging up to around 1050 nm with a contained energy of 202 mJ for 900 mJ pump and 2 mJ seed energy. Fig. 10(b) shows the simulated signal output energy of the 2nd NOPCPA stage as function of its seed energy. It exactly predicts the experimentally obtained energy value (gray dashed lines) and therefore implies that the experimental amplifier is operated close to its optimum. The same was observed for the parametric preamplifier [B4]. While the short wavelength edge is determined by the gain bandwidth, the NIR content is currently limited by the seed. Summing up, the experimental results for the present NOPCPA chain match the numerical predictions.

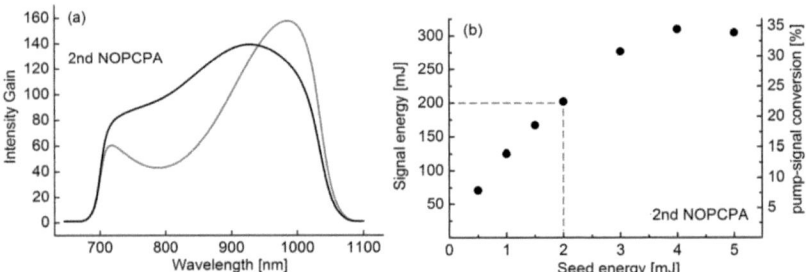

Fig. 10: (a) Simulated parametric intensity gain for the 2nd NOPCPA stage employing 900 mJ pump energy and 2 mJ (black), 5 mJ (red) seed energy. (b) Simulated signal energy and corresponding pump to signal energy conversion efficiency for the 2nd amplifier depending on its seed energy. The dashed lines label the experimental result, which matches the numerical prediction.

It should be noted that saturation depends on seed spectral energy density and phase-mismatch [74]. While some spectral components are still being amplified in the crystal, others can already exhibit back conversion to the pump. Fig. 10(b) also implies that the present parametric amplifier has not yet approached an overall spectral saturation but rather points to even higher output energies and energy conversion efficiencies around 35% in case of more available total pump energy.

Fig. 11(a) shows the numerically retrieved parametric signal phase shift for the preamplifier (gray) used in this work and the saturated 2nd NOPCPA stage with increasing degree of pump depletion (blue to red). The qualitative phase evolution as function of signal wavelength matches other numerical simulations in Ref. [66] and the analytic calculation shown in Fig. 6, which assumes negligible pump depletion. Fig. 11(a) points to increasing signal phase modulations during parametric amplification in prescence of pump depletion and should be taken into account for an accurate dispersion management especially for further energy-upscaling towards PW-scale few-cycle NOPCPA systems.

No contribution from AOPF could be detected experimentally through blocking the seed for the energy measurement. This is in contrast to other OPCPA systems, which employ a much weaker seed and work with higher parametric gain [33,80,84], yielding significant AOPF. Nevertheless, the AOPF should be regarded numerically for several reasons. First, the main parameters determining the AOPF level should be revealed for future OPCPA upscaling. Second, the high-dynamic range THG-AC described in section 3.3.2 allows for revealing contributions up to 10^{-11} contrast ratio with respect to the main pulse.

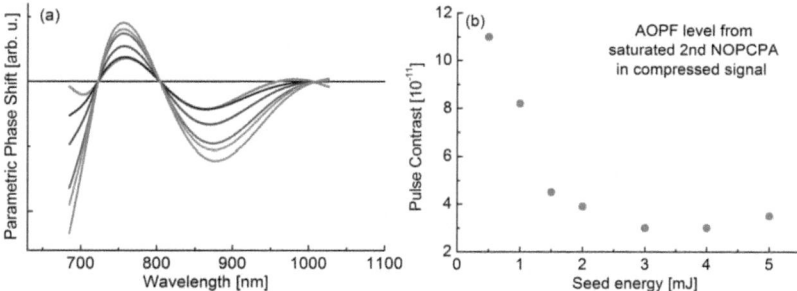

Fig. 11: (a) Simulated parametric phase shift applied to the signal during amplification for increasing (red: high) degree of pump depletion. (b) Simulated ps pulse contrast ratio after parametric amplification in the saturated NOPCPA stage including AOPF and compression of the main signal pulse to 8 fs.

Therefore, AOPF could directly be probed and the implementation of OPF in the numerical simulation could be judged. Fig. 11(b) shows an estimate for the level of AOPF after parametric amplification in the saturated NOPCPA stage just described above and compression of the stretched signal to 8 fs (compression ratio 2500). Saturation could allow for efficient built-up of AOPF. Some results for the first NOPCPA stage have already been shown in Ref. [B4]. The simulation assumes an initial stochastic field distribution of OPF in the 5 mm long BBO crystal, whose magnitude depends on the pump intensity and which competes with the signal for pump energy during parametric amplification at each nonlinear step. It turns out that an AOPF level of a few times 10^{-11} below the main compressed signal pulse is likely for the current LWS-20 design parameters. However, this contribution depends strongly on the seed energy of the NOPCPA stage as shown in Fig. 11. If the seed energy of the saturated NOPCPA stage drops below the 1 mJ level, the contrast ratio will decrease by about an order of magnitude. Approximately the same effect is observed for the parametric preamplifier and will limit the contrast ratio to about 10^{-10} if the seed energy from the frontend is reduced, e.g., by implementing XPW, which is supposed to enhance the seed contrast and cancel the ASE. Then AOPF will be detectable by the THG-AC and will limit the signal pulse contrast. However, even in this case the pulse contrast ratio provided by the chosen NOPCPA design will be superior to the high-energy CPA technique [20].

It is not straight forward to compress broadband light pulses as well as to characterize them. The temporal pulse shape and the phase information need to be determined to optimize the complex NOPCPA system towards Fourier-limited pulses and maximum peak power.

In many light-matter interactions it is required to provide an ultrashort light pulse without significant prepulses. Such prepulses can lead to sequences of excitations in molecular systems thereby reducing the time resolution and hinder a comprehensive data interpretation. In high-field experiments such as high-harmonic generation in overdense plasmas (solid state surfaces, [90,B10]) or particle acceleration from thin foils [91], prepulses or pedestals at about 10^{14} W/cm^2 would ionize the samples prior to the actual desired interaction thereby adversely affecting data interpretation.

An accurate pulse characterization requires a temporal and spectral resolution in the diagnostics much below the pulse duration and bandwidth for reliable data interpretation. Section 3.3 shows the result of LWS-20 pulse compression and contrast measurements using home-built devices.

3.3) Few-cycle Pulse Characterization

3.3.1) Pulse Compression

The spectral and temporal phase behaviour as well as the temporal pulse intensity have to be determined in order to gain understanding of the compressibility of the stretched, amplified and recompressed signal pulses amplified in NOPCPA systems. A single-shot, background- and dispersion-free second-order intensity AC and SHG-FROG were developed in collaboration with R. Tautz and used to characterize individual compressed sub-10 fs NOPCPA signal pulses (AC: at 10 Hz repetition rate) with 200 as resolution [B1,B2,B4,B6].

The development and the main results of the single-shot SHG autocorrelator and FROG are described in detail in the corresponding publication in Appendix B4, B1, B2:

Investigation of two-beam-pumped noncollinear optical parametric chirped-pulse amplification for the generation of few-cycle light pulses
Daniel Herrmann, Raphael Tautz, Franz Tavella, Ferenc Krausz, and Laszlo Veisz
Optics Express 18, 4170-4183 (2010).

Generation of sub-three-cycle, 16 TW light pulses by using noncollinear optical parametric chirped-pulse amplification
Daniel Herrmann, Laszlo Veisz, Raphael Tautz, Franz Tavella, Karl Schmid, Vladimir Pervak, and Ferenc Krausz
Optics Letters 34, 2459-24691 (2009).

Generation of 8 fs, 125 mJ Pulses Through Optical Parametric Chirped Pulse Amplification
Daniel Herrmann , Laszlo Veisz, Franz Tavella, Karl Schmid, Raphael Tautz, Alexander Buck, Vladimir Pervak and Ferenc Krausz
Advanced Solid-State Photonics (ASSP) 2009 paper WA3.

Here, due to the noncollinear interaction of the two fundamental beams (Fig. 12, red), all AC delays within a 600 fs time interval could be recorded in one shot. The detector read-out could be operated at 10 Hz. In this way, the shot-to-shot pulse duration stability and first indications of residual low-order phase terms could be obtained. To characterize the spectral and temporal phase of the signal pulse in more detail, the autocorrelator was augmented by a home-built SHG Frequency Resolved Optical Gating (SHG-FROG) device (Fig. 12), which was operated on single-shot basis with a shutter in the LWS-20 beam due to limited readout ability of the imaging spectrometer.

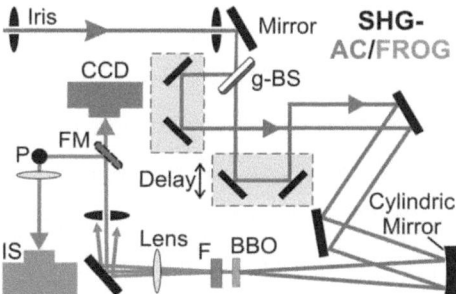

Fig. 12: Experimental layout of the combined single-shot SHG intensity autocorrelator (AC) and FROG device for the temporal and spectral characterization of sub-10 fs light pulses. g-BS: geometric beamsplitter, F: shortpass filter, FM: flipper mirror, P: beam 90°-rotating periscope, IS: imaging spectrometer and CCD line array.

In combination with the FROG inversion algorithm developed in the course of my Master thesis [92], the device provides a comprehensive knowledge of the generated few-cycle light pulses in the time and frequency domain. The retrieved information can then be compared with the result of a Fourier-limited signal pulse based on the experimental signal spectrum. The measured SHG AC and FROG traces have to be symmetric with respect to delay, due to the time inversion symmetry of the SHG signal:

$$S_{AC}(\tau) = \int_{-\infty}^{\infty} I(t)I(t-\tau)dt. \qquad (3.2)$$

This gives an experimental check if the obtained signal is a true and valid measurement. The experimental SHG-FROG trace involves a Fourier-transform of the SHG amplitude, thus spectrally resolving the SHG-AC:

$$I_{FROG}(\omega,\tau) = \left| \int_{-\infty}^{\infty} E(t)E(t-\tau)\exp(-i\omega t)dt \right|^2. \tag{3.3}$$

The SHG-FROG trace (Eq. (3.3)) is a pure real quantity and consequently contains no direct phase information. Hence, the bottom line of the FROG inversion algorithm is to determine the unknown phase information by solving the equation:

$$\sqrt{I_{FROG}(\omega,\tau)} \cdot \Phi(\omega,\tau) = \int_{-\infty}^{\infty} E(t)E(t-\tau)\exp(-i\omega t)dt, \tag{3.4}$$

where $\Phi(\omega,\tau)$ is an unknown complex phase function of unity magnitude. The iterative inversion algorithm needs to find the pulse electric field $E(t)_{guess}$, which corresponds to the minimal rms error (termed FROG error) per element of the spectrogram:

$$\varepsilon_{FROG} = \left[\frac{1}{N^2} \sum_{i,j=1}^{N} \left[I_{guess}(\omega_i,\tau_j) - I_{FROG}(\omega_i,\tau_j) \right]^2 \right]^{1/2}, \tag{3.5}$$

where N is the number of frequency and time points of a quadratic FROG trace. The inversion is based on the approach of the principle component generalized projections algorithm (PCGPA), which is particularly robust and fast for SHG-FROG [93].

It is not self-evident to perform an autocorrelation or a FROG measurement with sub-10 fs pulses. First of all, the entire setup has to be optimized for minimum dispersion (mirrors, beamsplitter, lenses). We only used reflective optics with negligible dispersion before the SHG in the autocorrelator. A D-shaped mirror was used to perform spatial beam splitting. Secondly, the type-I BBO crystal (cutting angle $\theta=28°$) for SHG has to be 5-10 μm thin to provide the required broadband phase-matching for few-cycle pulses at 800 nm central wavelength. A SHG phase-matching bandwidth ranging from about 500-1300 nm is achievable. Third, the unconverted fundamental and the individual parasitic SH beams have to be eliminated for undistorted measurements. This is achieved by inserting a shortpass filter (F) and a small iris behing the BBO crystal. Fourth, a proper imaging of the crystal plane onto the CCD line array (factor 4 magnification) and onto the entrance slit of the imaging spectrometer (factor 0.5 demagnification) has to be performed to achieve an adequate temporal δt (AC: ~200 as/pixel, FROG: ~500 as/pixel) and spectral δλ (0.3 nm/pixel) sampling of the unkown light pulse for both diagnostics.

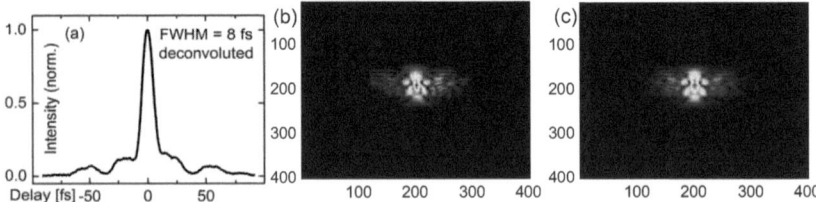

Fig. 13: (a) 2nd-order AC trace of the LWS-20 output pulse (deconvolution factor=1.35), (b) corresponding measured SHG-FROG trace and (c) the calculated FROG trace from the inversion result (τ: horizontal, ω: vertical).

The results of AC and FROG inversion are shown in Fig. 13 and Ref. [B1,B2]. A glass plate of known dispersion was inserted into the beam of the unknown pulse to remove the time ambiguity mentioned above. 7.9 fs, 130 mJ light pulses at 805 nm are compressed to within 5% of the Fourier-limit and could routinely be characterized in time and frequency domain. The FWHM intensity pulse duration corresponds to less than 3 optical cycles, while the electric field shows a FWHM duration of about 4 optical cycles. These few-cycle light pulses make the defined variation of the CEP phase interesting for light-matter interactions. The FROG inversion with an error of 0.6% revealed a fairly flat phase after adaptive pulse compression with phase 900 fs^2, -52500 fs^3, 71000 fs^4 applied by the AOM. The RMS of the pulse duration was determined to be 4% over hundreds of laser shots. To exclude strong dependencies of spatial beam quality for the measurements, several different beam fractions (containing min. 10 µJ) were selected for the measurements and no deviation of the result was found. Moreover, the same result was found for using the entire attenuated and downsized LWS-20 output beam.

Multi-shot pulse characterization shows undesired geometrical time smearing and assumes shot-to-shot identical pulses, which is both not the case for single-shot detection [92,93]. The FROG technique is in general superior compared to an AC due to the additionally gained phase information. It turned out that optimizing pulse compression using the SHG-FROG could improve the AC traces more than only via the AC technique. Furthermore, the SHG-FROG inversion algorithm can bring back trace information, which was lost during the measurement, e.g. because of clipping of the SH beam or detector imperfections. Due to its nonlinearity, FROG can reveal spatial chirp of the fundamental beam as shown in Ref. [92]. How sensitive it is becomes evident through comparison between the case of optimum compression Fig. 13 (900fs^2) and the raw FROG trace in Fig. 14 (700 fs^2), where mainly the 2nd-order phase was slightly readjusted via the AOM, yielding massive changes in the FROG trace.

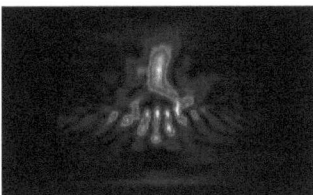

Fig. 14: Measured "octopus" SHG-FROG trace (τ: horizontal, ω: vertical) for slight deviation from the optimum setting for pulse compression.

3.3.2) Temporal Pulse Contrast

The variety of contributions to the contrast landscape of ultrashort pulses in high-power laser systems employing OPCPA and/or CPA asks for a long delay range (100 ps), high-dynamic range (10 orders of magnitude) cross- or autocorrelation measurement that can distinguish between pre- and post-pulses: ASE, AOPF, reflections, spectral modulation or clipping, uncompensated higher-order phase, pulse trains and pump noise transfer [94]. Furthermore, it was shown that postpulses can yield prepulses after pulse compression in case of a high B-integral in the amplifier chain [95].

The experimental results on the temporal pulse contrast are shortly described in the corresponding publications in Appendix B1, B3:

Generation of sub-three-cycle, 16 TW light pulses by using noncollinear optical parametric chirped-pulse amplification
Daniel Herrmann, Laszlo Veisz, Raphael Tautz, Franz Tavella, Karl Schmid, Vladimir Pervak, and Ferenc Krausz
Optics Letters 34, 2459-24691 (2009).

Generation of Three-cycle, 16 TW Light Pulses by use of Optical Parametric Chirped Pulse Amplification
Daniel Herrmann, Laszlo Veisz, Raphael Tautz, Alexander Buck, Franz Tavella, Karl Schmid, Vladimir Pervak, Michael Scharrer, Philip Russell and Ferenc Krausz
The European Conference on Lasers and Electro-Optics (CLEO/Europe) 2009 paper CG1_3.

Third-harmonic generation (THG) on interfaces between transparent optical media [96] or noncollinear THG by using two subsequent nonlinear crystals have been employed [97]. In this work we used and upgraded an existing THG scanning autocorrelator initially developed by K. Schmid [98]. The device alignment was changed and it was extended by thicker wedged BBO crystals to achieve 11 orders of magnitude detection sensitivity [B1], which is better than other current state-of-the art devices [34,79]. The autocorrelator was developed for ultrafast pulse contrast measurements with a scanning range of more than 600 ps and is based on cascaded type-I phase-matched SHG and SFG in BBO crystals yielding the background-free intensity third-order correlation function:

$$S_{THG}(\tau) = \int_{-\infty}^{\infty} I^2(t) I(t-\tau) dt. \tag{3.6}$$

In a third-order correlation, every pre-pulse (post-pulse) will generate a peak at positive (negative, before the main pulse) delays as well, but this artifact will be of smaller amplitude than the peak at the real position of the satellite. In a third-order correlation-trace, the peak value of the actual satellite pulse is approximately the square of the peak-value of the artifact. In another view, if a correlation-trace is normalized to one, the contrast ratio between the main pulse and the artifact pulse is given by about the square of the true contrast of the satellite.

The optical layout and a detailed description of the THG AC is given in Ref. [98]. To reach a dynamic range exceeding 10 orders of magnitude, several device aspects were changed and optimized. The beam sizes have been reduced and the crystal lengths were increased from 100 µm to 200 µm (SHG) and 400 µm (SFG) yielding higher SHG and THG signals. Accordingly, the phase-matched pulse bandwidth decreased but was still sufficient to provide phase-matching around 800 nm fundamental wavelength, where the mayor contrast contributions of ASE from the Ti:Sa frontend and of OPF from the NOPCPA stages would be located. Nevertheless, the crystal angles were also detuned to achieve phase-matching at various central wavelengths and no additional contribution to the LWS-20 pulse contrast could be found.

Fig. 15 shows the recorded pulse contrast after compression of the Ti:Sa CPA frontend (25 fs, 0.8 mJ, Femtolasers GmbH) and the LWS-20 NOPCPA system (8 fs, 130 mJ). In general, the scanning range was chosen such to first measure along a wide delay range (-80 to +80 ps, long range) and then to check in a short range (-20 to +20 ps) wether more features can be found with higher time resolution. Afterwards the contrast curves are merged (Fig. 15). By blocking the fundamental, the SH and the TH individually, the contrast detection limit could be determined as 10^{-10} to 10^{-11} (Fig. 15, [B1]).

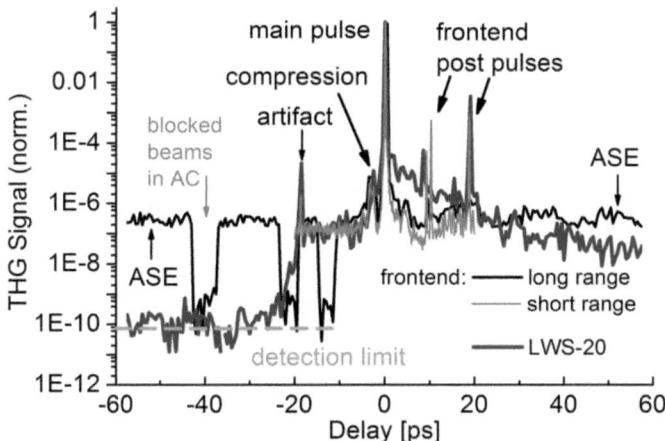

Fig. 15: High dynamic range THG-AC to determine the pulse contrast of the output pulse of the CPA frontend (black: long delay range, red: short range with higher time resolution) and the LWS-20 NOPCPA system (blue curve, long+short range merged). To determine the detection limit (orange dashes), the individual beams in the THG-AC are blocked separately.

Fig. 15 reveals several contributions to the normalized pulse contrast of the CPA frontend: (i) a long-lived contrast degrading ASE pedestal at about 5×10^{-7} before and after the main pulse, (ii) 3 post-pulses probably originating from reflections in the commercial multi-pass amplifier all at about 10^{-3} and (iii) a contrast degrading hump just before the main pulse most likely due to incomplete compression (higher-order residual phase, SPM) in the hybrid prism-dispersive mirror compressor.

It turns out that the pulse contrast of the LWS-20 NOPCPA system is strongly limited by these contrast contributions of the CPA frontend laser system: (i) the ASE pedestal at 800 nm is further amplified within the pump pulse duration (78 ps FWHM) in the NOPCPA chain. The onset of this contrast degrading contribution depends on the pump-seed delay chosen in the parametric amplifiers and is at about -20 ps (Fig. 15) for the optimum signal pulse energy and spectrum. Its magnitude can be minimized to around 10^{-7}-10^{-8} and increases with parametric gain and degree of saturation in the NOPCPA stages. This pedestal cannot be due to AOPF generated in the NOPCPA stages, because its level is too high as stated by the above shown numerical simulations (Fig. 11) and first efforts undertaken to clean the frontend pulse contrast via cross-polarized wave generation (XPW, [99]) with a result of concurrent suppression of this pedestal in the LWS-20 pulse [B3]. The fluctuating signal before -30 ps is originating from back ground noise at the detection sensitivity. (ii) The post-pulses at about +19 ps originating from the multi-pass amplifier generate visible artifacts (at -19 ps) at the expected signal intensity and fall outside the pulse duration (40 ps group delay) of the main stretched seed. They therefore can extract more pump energy as the post-pulse at about +10 ps, which has to compete with the main pulse. This is apparent from the by 2 orders of magnitude different intensity level of these post-pulses at the LWS-20 output, while at this low intensity level their contribution to the measured LWS-20 output energy is negligible. This was also verified after implementing XPW into the frontend, where similar signal energies could be obtained at about the same pump energies in the NOPCPA chain [B3]. (iii) The hump at -3 ps is possibly mainly due to higher-order residual phase in the complex LWS-20 dispersion management (grism stretcher and hybrid glass-dispersive mirror compressor) and is about the magnitude as the hump at the same delay in the CPA output pulse contrast.

Overall, the designed NOPCPA chain consisting of a relatively (compared to other seed generation schemes) strong few-µJ seed, a parametric preamplifier and a saturated parametric amplifier (both operated at moderate gain) yield a seed-limited pulse contrast after stretching, parametric amplification and compression. No contrast-degrading contributions from AOPF can be detected within 10 orders of magnitude dynamic range as a result of the chosen NOPCPA parameters. This is confirmed by numerical simulations (Fig. 11) and contrast measurements applying XPW to the seed pulse, where AOPF contributions are not expected above the 10^{-10} level at this chosen seed energy. Consequently, the present LWS-20 NOPCPA system can in principle provide a high ($\geq$ 10 orders of magnitude) contrast ratio at a very few ps before the main pulse. Nevertheless, the pulse contrast shown in Fig. 15 and Ref. [B1] is already higher as in many other high-energy laser systems providing light pulses below 20 fs duration [20,29,34] and is sufficient for many laser-plasma interactions.

Chapter 4: Investigation of Two-beam Pumped NOPCPA

In section 2.1 it was shown that matching of the signal group velocity with the idler group velocity in signal direction is necessary to eliminate the first-order derivative of the wavevector-mismatch in Eq. (2.6) and can be achieved in NOPCPA. In chapter 3, the potential of the NOPCPA concept using one pump beam was outlined through the generation of sub-8 fs light pulses containing more than 130 mJ with the potential of energy-upscaling. To achieve an even broader phase-matching plateau around the central seed wavelength at small wavevector mismatch, the second-order wavevector-mismatch in Eq. (2.6) also needs to vanish. Therefore, an additional degree of freedom is required. Apart from the option to employ angular dispersion or a broadband chirped pump pulse [6,100], angular detuning of the nonlinear crystal in multiple passes or multiple amplification stages was suggested [101]. Alternatively, using two pump beams to amplify adjacent spectral components of the seed bandwidth in the same nonlinear crystal (*TBP-NOPCPA*) was suggested and investigated for shortening from 98 to 61 fs [102].

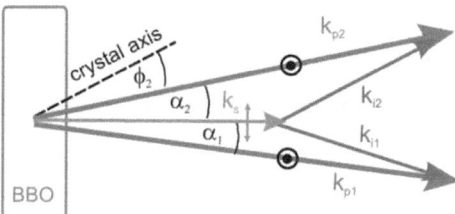

Fig. 16: Experimental broadband phase-matching geometry for NOPCPA employing a type-I BBO crystal pumped in the φ-plane by two beams at 532 nm. Both pump pulses experience the same phase-matching angle θ_{pm} but possess individual azimuthal angles $\phi_{1,2}$ and internal noncollinear angles $\alpha_{1,2}$ yielding two separate idler beams.

In this thesis work [B4,B6], the potential of this concept is studied in detail for few-cycle pulses and the reduction of the pulse duration from 8 to 7 fs is shown for similar phase-matching as used for LWS-20. Fig. 16 shows the experimental broadband phase-matching geometry for TBP-NOPCPA employing a 5 mm long type-I BBO crystal pumped in the φ-plane by two pulses of 6 mJ each at 532 nm with a FWHM pulse duration of 78 ps. Both pump pulses experience the same phase-matching angle θ_{pm} but possess individual azimuthal angles $\phi_{1,2}$ as well as different internal noncollinear angles $\alpha_{1,2}$ (calculated values here: $\alpha_1=2.16°$, $\alpha_2=2.22°$, $\theta_{pm}=23.62°$) yielding two separate idler beams. Seed source, stretcher-compressor arrangement and single-shot 2nd-order autocorrelator from chapter 3 are used. Fig. 17(a) shows that up to 30 nm signal bandwidth below 700 nm can be added to the signal spectrum from Fig. 9 employing TBP, while the pump to signal conversion efficiencies for TBP are similar to using one pump beam with the same total energy. Interference patterns between the two pump beams of the same polarization orientation are detected and are not transferred to the common signal beam.

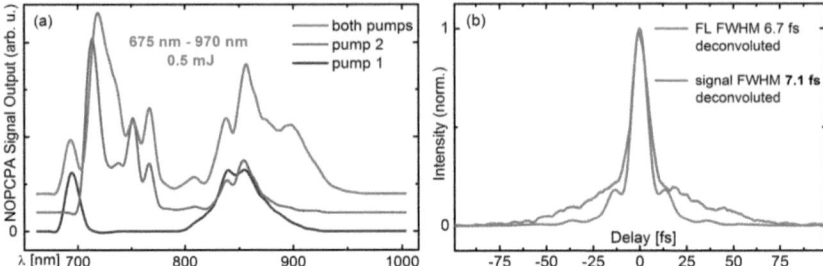

Fig. 17: (a) Spectral energy density of the signal amplified only by pump 1 (blue), only by pump 2 (green) and using both pumps (red) along with the measured AC trace (b).

Adaptive pulse compression to within 6% of the Fourier-limit (6.7 fs) yields 7.1 fs (FWHM) light pulses containing 0.35 mJ of energy in the NIR spectral range at Fig. 17This indicates that compression of signal spectra composed by TBP-NOPCPA in the same crystal is possible. , the numerical NOPCPA simulations including the nonlinear coupled wave equations and AOPF from section 2.3 are extended for the case of TBP. The gain curve reproduces the amplified signal spectrum and the AOPF level for using one pump seems to be fairly unaffected by splitting the pump energy into two pulses [B4]. Consequently, TBP-NOPCPA is a promising approach to extend the phase-matching bandwidth even in the few-cycle regime with the prospect to yield mJ-level signal energies.

All experimental and theoretical results of chapter 4 are described in detail in the corresponding publication in Appendix B4, B6:

Investigation of two-beam-pumped noncollinear optical parametric chirped-pulse amplification for the generation of few-cycle light pulses
Daniel Herrmann, Raphael Tautz, Franz Tavella, Ferenc Krausz, and Laszlo Veisz
Optics Express 18, 4170-4183 (2010).

Approaching the Full Octave: Noncollinear Optical Parametric Chirped Pulse Amplification with Two-Color Pumping
Daniel Herrmann, Christian Homann, Raphael Tautz, Laszlo Veisz, Ferenc Krausz, and Eberhard Riedle
Advanced Solid-State Photonics (ASSP) 2011 paper JWC1.

Chapter 5: Approaching the full Octave via Two-color Pumping

Apart from the strategies discussed in chapter 4 for an enhancement of the amplified signal bandwidth beyond the optimum case of one-beam pumped NOPCPA (LWS-20, chapter 3), in this chapter we discuss the idea to pump two cascaded type-I NOPCPA stages at different pump wavelengths. In particular, phase-matching calculations shown in Fig. 18 imply a significantly increased phase-matching bandwidth approaching an octave in the Vis-NIR spectral region using the second- (SH, 2ω) and third-harmonic (TH, 3ω) of a Nd-YAG pump laser. In this approach, the UV pump amplifies the visible (yellow-red) part of the light spectrum adjacent to the broadband NIR spectral range amplified by the green pump. The calculations in Fig. 18 are performed according to the discussion in section 2.1.

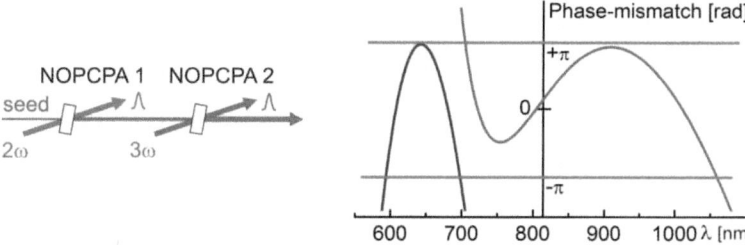

Fig. 18: By pumping the first stage of a cascaded NOPCPA chain by the SH (green) of the Nd:YAG pump laser and the second stage by the TH (blue), the phase-matching bandwidth can be extended approaching an octave (580-1050 nm). The effective phase-mismatch as a function of seed wavelength including wave-vector mismatch ΔkL and parametric signal phase shift is calculated for the individual NOPCPA stages: λ_p=532 nm, θ_{pm}=23.62°, α=2.23°, L=5 mm (green) and λ_p=354.7 nm, θ_{pm}=34.58°, α=3.40°, L=3 mm (blue). The red horizontal lines label ±π.

The experimental and theoretical results of chapter 5 are described in detail in the corresponding publication in Appendix B5, B6:

Approaching the full octave: Noncollinear optical parametric chirped pulse amplification with two-color pumping
D. Herrmann, C. Homann, R. Tautz, M. Scharrer, P. St.J. Russell, F. Krausz, L. Veisz, and E. Riedle
Optics Express 18, 18752-18762 (2010).

Approaching the Full Octave: Noncollinear Optical Parametric Chirped Pulse Amplification with Two-ColorPumping
Daniel Herrmann, Christian Homann, Raphael Tautz, Laszlo Veisz, Ferenc Krausz, and Eberhard Riedle
Advanced Solid-State Photonics (ASSP) 2011 paper JWC1.

Fig. 18 shows the experimental scheme for two-color pumped NOPCPA. Employing tens of mJ pump energy at the SH and TH of a Nd:YAG pump laser (78 ps FWHM, 10 Hz) to pump a first NOPCPA stage consisting of a 5 mm long type-I BBO crystal and a second NOPCPA stage consisting of a 3 mm long type-I BBO crystal. Broadband parametric amplification is observed from 575-1050 nm approaching an octave. The spectral energy density shown in Fig. 19 allows for a Fourier-limit of 4.5 fs (inset), which corresponds to just two optical cycles at the central wavelength of 782 nm. Fig. 19 also reveals that parametric amplifiction is achieved from 700 to 1050 nm in the 1st NOPCPA stage, while the gain bandwidth ranges from 575 to 740 nm in the 2nd NOPCPA stage. These results match the predicted signal bandwidth for the individual NOPCPA stages based on the phase-mismatch analysis shown in Fig. 18. The UV pumped amplifier even slightly exceeds its expected signal bandwidth.

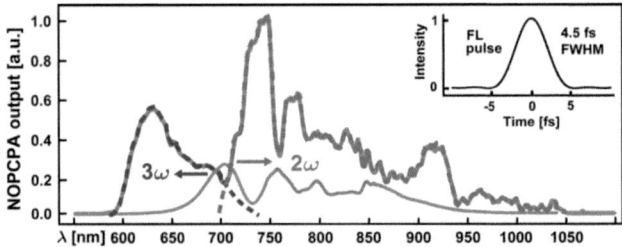

Fig. 19: By cascaded two-color pumped NOPCPA, the spectral energy density of the signal pulse (575–1050 nm, red solid curve) allows for a Fourier-limit of 4.5 fs (inset). This spectrum is composed of the spectral region amplified only by the SH (700–1050 nm, green dashed curve) and the TH alone (575–740 nm, blue dashed curve). The seed spectrum (direct T-HCF output, not to scale) is shown as gray solid curve.

For this experiment, the very broad seed was stretched using glass and hence did not allow for a pulse compression with the available techniques. This was purely due to the short term availability of components and can be avoided in future work. By variation of the group delay between the seed spectral boundaries, the optimum pump to signal energy conversion efficiency of about 13% was achieved for a GD of 69 ps while still maintaining the signal bandwidth. A total signal energy of about 3 mJ was measured. The GD for 700 to 1020 nm and for 575 to 740 nm is roughly 33 ps and 43 ps, respectively. In this case, the conversion efficiency is about 14% and 9% in the NOPCPA stages pumped selectively by the SH and the TH, which is the highest conversion efficiency obtained for the SH-pumped stage. This proofs that the 40 ps seed GD chosen for case of the LWS-20 is close to the optimum.

To study the phase behavior in the region of spectral overlap and judge if such composed spectra can be compressed, a proof-of-principle experiment was performed by extending two-color pumping also for a 100 kHz NOPA systems employing a Yb:KYW laser (300 fs FWHM, 100 kHz). The seed was generated in a 2 mm thin YAG plate to minimize dispersion.

Few-µJ pumping leads to about 1 µJ total signal energy after amplification in 2-3 mm type-I BBO crystals with the SH ($\theta_{pm}=24°$, $\alpha=2.3°$) and TH ($\theta_{pm}=35.5°$, $\alpha=2.7°$) in cascaded NOPA stages. Due to the shortness of the pump pulses, the signal bandwidth was not limited by the phase-mismatch in the crystals.

In this low-energy cascaded NOPA system, focusing of the pump and seed beams towards the BBO crystals is required to achieve adequate intensities (pump: ca. 200 GW/cm^2) for parametric amplification. We find that careful matching of seed and pump in both NOPA stages is desired to obtain homogeneous spatial profiles of the signal. Interestingly, we observe that spatial overlap between the amplified signal and the unamplified seed in a NOPA stage is not ensured automatically. This spatial overlap is a prerequisite for amplification of neighboring spectral regions of the same beam in the two cascaded NOPA stages. If the overlap is not fulfilled two signal beams with spectrally disjunct spectral energy density will result apart from the unamplified seed and all these beams will propagate along their individual optical path. We observe that the amplified signal usually tends to be steared towards the pump direction and hence leaving the BBO crystal between pump and unamplified seed. However, we find that only in case of the phase-matching geometry compensating for Poynting vector walk-off (Fig. 5) and therefore also (partially) compensating the spatial separation between seed and signal, the spatial overlap between seed and signal can be achieved. In this case the pump polarization lies in the plane of noncollinear interaction defined by pump and seed. Since the two NOPA stages are pumped at two different pump wavelengths, the pump spatial properties inherently differ and make a proper matching of pump-seed divergence and signal-seed spatial overlap challenging and might involve slight deviations from the optimum phase-matching parameters.

The gain bandwidth of the individual amplifiers overlap by purpose at around 700 nm, as in the high-energy case in Fig. 19. By using a prism compressor and 2nd-order autocorrelation, the signal pulse from the green-pumped amplifier is compressed to 23 fs close to its Fourier-limit of 20 fs with the prism compressor. If the blue pump is then unblocked, the pulse duration is about a factor of 2 reduced by adding the spectral components in the blue-pumped amplifier to the signal pulse amplified in the green-pumped 1st NOPA stage [B5]. The composite signal pulse amplified in both NOPA stages is compressed to 13 fs close to its Fourier-limit of 11 fs. One can draw the conclusion from this finding that a signal spectrum composed by cascaded two-color pumped NOPCPA can be compressed close to its inherent limits of about two optical cycles. **Experimental evidence for parametric phase.**

However, a few important experimental challenges have to be overcome. Actually, the signal pulse only amplified by the blue-pumped NOPA is fairly long (95 fs) for this prism compressor setting and can be compressed close to its Fourier-limit (28 fs) for a significant reduction of negative group delay dispersion (GDD) in the compressor (further insertion of 2nd prism). Despite the fact that the signal pulses from the individual NOPA stages experience the same material dispersion throughout the NOPA system, significantly differing phase applied in the compressor is needed to achieve pulse compression close to the Fourier-limit. Consequently, a phase dependent on the parametric amplification has to come into play.

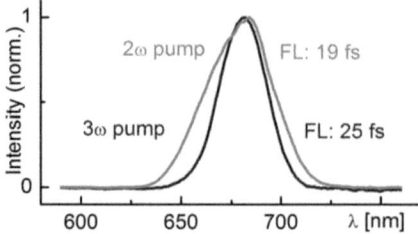

Fig. 20: Normalized spectral energy density of the signal amplified in the green-pumped (green) and blue-pumped (blue) NOPA stage centered at 670 nm along with their Fourier-limits.

To test this hypothesis, both NOPA stages were tuned to amplify almost the same signal spectrum centered at 670 nm as shown in Fig. 20 along with the Fourier-limits around 20 fs. The signal pulses amplified in the individual NOPA stages are then compressed separately. If the prism compressor is adjusted to achieve compression of the signal from the green-pumped NOPA close to its Fourier-limit, the signal amplified from the blue-pumped NOPA is then way off its own Fourier-limit as shown in Fig. 21(a). By significantly reducing the negative GDD in the prism compressor, this observation can be reversed such that the signal from the blue-pumped stage is compressed in contrast to the signal from the green-pumped NOPA as shown in Fig. 21(b).

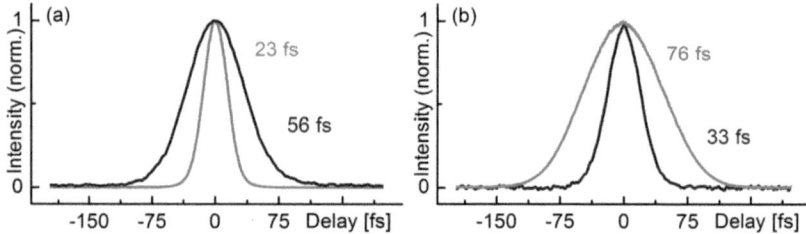

Fig. 21: Normalized 2nd-order AC traces along with the deconvoluted FWHM pulse durations. (a) If the signal from the green-pumped NOPA (green) is compressed close to its Fourier-limit, the signal from the blue-pumped NOPA (blue) is uncompressed. (b) By significantly reducing the negative GDD in the prism compressor, this situation can be the opposite.

This implies that the signal experiencing parametric amplification in the blue-pumped NOPA exhibits less positive second-order phase (positive temporal chirp) as the signal amplified in the green-pumped NOPA. If one performs the calculations of the parametric phase imprinted on the signal during amplification according to Eq. (2.4) as shown in Fig. 22 for the experimental NOPA stages pumped by the SH (green) and the TH (blue) of the pump laser, the cascaded parametric amplifiers then show opposing signs for this parametric phase. While for the green-pumped NOPA a parabola with a negative 2nd-order derivative in the frequency domain results, the blue-pumped NOPA leads to a correspondingly positive sign. Therefore, parametric amplification in the first green-pumped NOPA leads to an additional positive temporal chirp via Fourier-transformation, which can be partially compensated in the second blue-pumped NOPA yielding the observed different compressor settings for pulse compression of the individual signals separately. The signal amplified in the blue-pumped NOPA needs less negative GDD in the prism compressor, since part of this compensating phase is already provided by the parametric amplification.

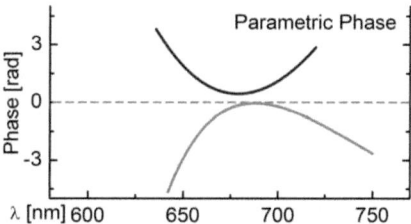

Fig. 22: Calculated parametric phase according to Eq. (2.4) imprinted on the signal during amplification for the experimental NOPA stages pumped by the SH (green) and the TH (blue) of the pump laser. The corresponding spectra are shown in Fig. 20.

These observations are clear experimental evidence for the existence of a parametric phase imprinted on the signal during parametric amplification and its consequences in the frequency and space domain. The phase magnitude and sign is strongly pump wavelength-dependent and can possess a positive or negative component of GDD as shown in Fig. 22. These findings are in good agreement with the theoretical discussion in course of section 2.1 and Ref. [B4,B6]: The parametric phase depends in particular on wave-vector mismatch [66], pump intensity and employed pump wavelength. The first two parameters are not too different for the present experimental NOPA stages, hence the pump wavelength is expected to yield the significant parametric phase effect. The insight gained from numerical simulations shown in Fig. 11 point to even more significant spectral and spatial influence of this parametric phase in case of pump depletion and saturation. Even for this low-energy NOPA system employing one pair of two-color pumped cascaded amplifiers, the spectral influence was observed in terms of challenging pulse compression and the spatial equivalents such as difficult seed-signal beam overlap. It can be expected that the spectral and spatial effects observed in the present experiment also affect larger NOPCPA/NOPA systems with multiple cascaded amplifiers eventually reaching saturation, although a certain regime of two-stage high-energy NOPCPA in this thesis work could provide high-quality output and was shown to be scalable in energy to a certain extend [B1,35]. The parametric phase effects will be pronounced when several pump wavelengths are employed. Moreover, the results of this thesis and Ref. [66] clearly show that – in contrast to general assumptions [67] - the CEP can in general not be assumed to remain unchanged during parametric amplification.

Chapter 6: Applications of Multi-TW, Few-cycle Light Pulses

6.1) Few-cycle Laser-Driven Electron Acceleration

The experimental results in this section are based on the publications:

Few-Cycle Laser-Driven Electron Acceleration
K. Schmid, L. Veisz, F. Tavella, S. Benavides, R. Tautz, D. Herrmann, A. Buck, B. Hidding,
A. Marcinkevicius, U. Schramm, M. Geissler, J. Meyer-ter-Vehn, D. Habs, and F. Krausz
Physical Review Letters 102, 124801 (2009).

Density-transition based electron injector for laser driven wakefield accelerators
K. Schmid, A. Buck, C. M. S. Sears, J. M. Mikhailova, R. Tautz, D. Herrmann, M. Geissler,
F. Krausz, and L. Veisz
Physical Review Special Topics-Accelerators and Beams 13, 091301 (2010).

Few-cycle laser-driven acceleration of electrons in a laser-induced plasma, the so-called laser wakefield acceleration (LWFA), has been a subject of interest for the past decades that relies on strongly driven plasma waves [103-105]. This approach allows for the generation of accelerating gradients in the range of several 100 GV/m, which is 3-4 orders of magnitude larger than the fields attainably by conventional RF-accelerators, which are mainly limited by electrical breakdown in the accelerating structures to approximately 100 MV/m. Since the acceleration length required for achieving a particular electron energy is inversely proportional to the accelerating field, this attribute leads also to a drastic reduction of the size and cost of the accelerator facility. Moreover, the specific characteristics of electron bunches accelerated by ultrashort light pulses, i.e. the pulse duration of a few fs, the high charge density and the high brilliance, open up new possibilities in many applications of these electron beams such as research employing free-electron lasers (FEL) or deployment in clinical environments.

An intense spatially varying light pulse drives a plasma wake through the action of the nonlinear ponderomotive force. The condition for laser-light propagation in plasma is satisfied in an underdense plasma, where the plasma frequency

$$\omega_p = \sqrt{\frac{n_e e^2}{\varepsilon_0 m_e}} \qquad (6.1)$$

is smaller than the frequency of the light (n_e: electron density). In this case the refractive index of the plasma

$$\eta = \sqrt{1 - \frac{\omega_p^2}{\omega^2}} \qquad (6.2)$$

is real and smaller than one, leading to group velocities $c \cdot \eta$ smaller and phase-velocities c/η larger than the speed of light c. The electron critical density separating the underdense regime

Chapter 6: Applications of Multi-TW, Few-cycle Light Pulses

from the overdense plasmas is about $1.74 \cdot 10^{21}$ cm^{-3} for light at 800 nm. For comparison, completely ionized He at 1 bar would lead only to about 3% of the critical free electron density, while solid state materials offer electron densities of 10^{22}-10^{23} cm^{-3} and are hence used for overdense plasmas. The cycle-averaged kinetic energy of an electron in an electromagnetic field is given by the pondoromotive potential

$$\Phi_{pond}(\vec{r}) = \frac{e^2}{4m_e\omega_L^2} E(\vec{r})^2 \qquad (6.3)$$

which is linked to the light pulse intensity via $I_L(\vec{r}) = \langle |S_L| \rangle_T = \frac{1}{2}c\varepsilon_0 E(\vec{r})^2$.

Hence, the spatial variation of the light pulse electric field leads to the pondoromotive force

$$\vec{F}_{pond} = -\vec{\nabla}\Phi_{pond} \qquad (6.4)$$

pushing the electrons away from regions of high light intensity towards those of low intensity. Electrons are then attracted by the remaining heavier positively charged ions and for a short light pulse even get pushed backwards by $\vec{F}_{pond}$ within the plasma wave and thereby increasing the amplitude, which travels with the group velocity of the light pulse. The plasma wave is maximized for a FWHM pulse duration matching half the plasma wavelength (typically 5-20 µm). In the non-relativistic case, the electrons will only be accelerated transversally out of the focal region, while at relativistic intensities (>10^{18} W/cm^2) the relativistic drift motion starts to come into play causing the electrons to be directed forwards and sideways (pondoromotive scattering) with a figure-of-eight motion in the co-moving frame. Focusing an intense laser pulse to a finite spot size into a plasma therefore can lead to laser accelerated electrons.

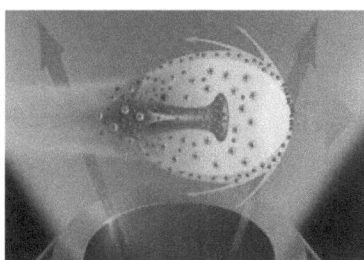
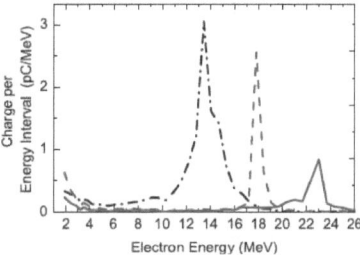

Fig. 23: Schematic of LWFA in the bubble regime (left) focusing a few-cycle light pulse (white) with relativistic intensities into an underdense plasma (blue). Electrons (red) are transversally pushed away and are then attracted by the heavy ions (blue) while some of the electrons are captured by the bubble where they are accelerated. Experimental single-shot electron spectra (right) for using multi-TW few-cycle light pulses from the LWS NOPCPA system taken from Ref. [B8].

Numerical studies using Particle In Cell-codes (PIC) have predicted for a certain parameter regime the emergence of monoenergetic electron bunches with relativistic energies and few-fs duration from laser excited plasma waves [106]. In this nonlinear regime, the plasma waves are strongly driven by a laser pulse of highly relativistic intensity ($\sim 10^{19}$ W/cm^2) and a duration and diameter that are matched to the plasma wavelength. The principle of this acceleration scheme is shown in Fig. 23. The ponderomotive force of the light pulse pushes free electrons transversally out of its path leaving the positively charged and much heavier ions behind. The electrons are pulled back to the axis by the electrical field created by charge separation after a propagation length comparable to a plasma wavelength. According to the simulations, a fraction of the returning electrons are trapped in a cavity trailing the laser pulse, termed a "bubble" (the bubble radius is about half the plasma wavelength), and accelerated by the strong longitudinal electric field, producing relativistic electron bunches with narrow-band energy spectra and small Maxwellian background. The bubble constitutes a highly anharmonic Langmuir wave that breaks down after the first period of its oscillation leading to self-trapping of electrons in the potential of the wave bucket, which are then accelerated as a result of energy transfer from the light pulse to the relativistic electron beam.

Sub-10 fs light pulses thereby directly excite a strong plasma wave even with few-10 mJ of energy while slightly longer (few-10s of fs) pulses need to get shortened to excite that strongly and have to contain J-level energy. The shortening is done by nonlinear self-modulation with the plasma period in an extended plasma channel. Instabilities grow during the electron acceleration process due to this mechanism. Only the leading part of the pulse contributes, which leads in this case to less efficient energy transfer between light wave and electrons. For light pulse durations much longer than the plasma period, direct laser acceleration (DLA) is considered, where electrons receive most of their energy directly from the transverse laser field leading to a predominant movement of electrons in the laser direction via the $\vec{v} \times \vec{B}$-force and broad thermal-like exponential spectra.

The scaling laws for LWFA in the bubble regime are still debated but theoretical models predict for the maximum electron energy E_{mono} of the quasi-monoenergetic peak and for the number of accelerated electrons within that peak N_{mono}:

$$E_{mono} \propto \frac{\sqrt{E_L \cdot \tau}}{\lambda}, \quad N_{mono} \propto \sqrt{E_L}, \quad (6.5)$$

where λ is the laser central wavelength, τ the $1/e^2$ pulse duration and E_L the laser pulse energy.

The energy scaling law is so far confirmed by our LWFA experiments performed with the LWS NOPCPA system. First experiments focusing 40 mJ, 8 fs light pulses at 800 nm central wavelength focused down to 6 µm FWHM diameter ($1.2 \cdot 10^{19}$ W/cm^2) into a He gas jet yielded up to 25 MeV quasi-monoenergetic (<10% relative energy spread) electron beams containing 1-10 pC of charge ($\sim 10^7$ electrons) as shown in Fig. 23 [B8]. Later, with an increased pulse energy of 65 mJ, electron energies exceeding 30 MeV were obtained [B9]. However, the stability and reproducibility of this nonlinear process remains to be improved and the measured charge is up to two orders of magnitude below the predicted value for these laser parameters. For the currently available LWS-20 output parameters (>130 mJ, 7.9 fs at 805 nm), the scaling laws in Eq. (6.5) would predict electron energies readily exceeding 50 MeV and accelerated charges of more than 500 pC.

6.2) High-harmonic Generation from Solid Surfaces

Most of the experimental and theoretical results in this section are based on the following publication, while some remain so far unpuplished:

Toward single attosecond pulses using harmonic emission from solid-density plasmas
P. Heissler, R. Hörlein, M. Stafe, J.M. Mikhailova, Y. Nomura, D. Herrmann, R. Tautz, S.G. Rykovanov, I.B. Földes, K. Varjú, F. Tavella, A. Marcinkevicius, F. Krausz, L. Veisz, G.D. Tsakiris
Applied Physics B 101, 511–521 (2010).

The generation of coherent high-order harmonics from the interaction of ultra-intense femtosecond laser pulses with solid density plasmas (also termed *Surface High-harmonic Generation* – SHHG [90]) holds the promise for table-top sources of intense extreme ultraviolet (XUV) and soft x-ray (SXR) radiation. Since SHHG has no inherent limit on the employed laser intensity (in contrast to gas harmonics sources), they give rise to the prospect of combining attosecond pulse duration with the photon flux currently only available from large-scale FEL or synchrotron facilities. The development of ultra-intense sources of single as-pulses from the interaction of intense laser pulses with solid surfaces could advance making XUV-pump/XUV-probe investigations of fundamental processes with attosecond time resolution feasible in the future, while for some envisioned applications single-as pulses are desired [3]. Using intense multi-cycle light pulses, it was shown that the individual harmonics are phase-locked and an as pulse train could be characterized [107]. For the ultimate goal, few-cycle light pulses with controllable CEP as driver of the light-matter interaction are required [108].

As a first step towards this ambitious goal, the 16 TW, sub-three-cycle light pulses generated by LWS-20 are focused to relativistic intensities of about $2 \cdot 10^{19}$ W/cm^2 onto a fused silica disc under 45° angle of incidence. The harmonic radiation was then recorded with an XUV imaging spectrometer as as shown in Fig. 24.

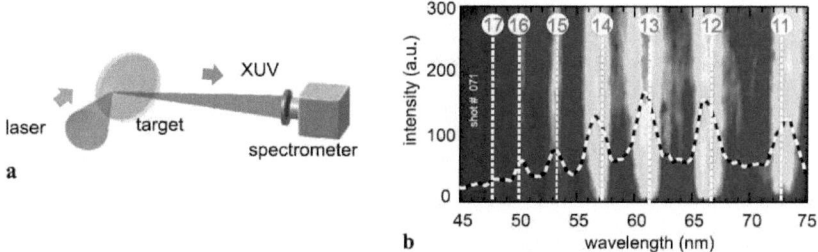

Fig. 24: (a) Schematic layout of the experimental setup, (b) Measured single-shot XUV spectrum with labeled harmonic number (taken from Ref. [B10]).

The dominant SHHG mechanism under investigation here is called *Coherent Wake Emission* (CWE) [90]. According to a 3-step model [B10], the incident electric field component perpendicular to the target of the obliquely incident, p-polarized ultrashort light pulse launches electrons into the vacuum. Some of these energetic electrons return back to the plasma in the second half-cycle of the period forming bunches in the plasma. This yields resonantly driven plasma oscillations where an integer multiple of the driving frequency ω_L matches the local plasma frequency ω_p ($q \cdot \omega_L = \omega_p$), while each electron moves in the combined electric field of the incident light pulse and of the space charge. Phase matching results in periodic bursts of radiation at the local plasma frequency containing harmonics up to the maximum electron density n_e.

The origin for the electron bunching is mainly the electron dynamics for different phases of the sinusoidal laser electric field. At the beginning of each positive-to-negative field crossing some electrons are revealed but due to the weak field at the outer part of the light pulse their excursion is short. As the field continues to grow, the electrons move further from the surface but due to their higher velocity they return about at the same time with the previous ones. This results in the formation of electron bunches. The harmonic emission contributing to the formation of as pulses coincides with the return of the revealed electrons, while the electron bunches are not equidistant as a result of the intensity envelope of the incident pulse.

The performed experiments show that a s-polarized or elliptically polarized incident electric field leads to reduced XUV intensity, thereby pointing to polarization gating as tool to generate single as pulses. This effect is likely due to the reduced electric field component in the plane of incidence and a variation of angles under which the electrons reenter the plasma reducing electron bunching. This outcome also shows that the temporal pulse contrast shown in Fig. 15 is outstandingly high, preserving a clean interaction surface, because no clear picture of the CWE dependence on incident electric field polarization could be obtained with other laser systems before.

Moreover, the experimental outcome is likely to reveal the *few-cycle nature* of the harmonic emission. High-harmonics of the 805 nm fundamental wavelength were recorded up to the 20th harmonic, the highest possible harmonic for CWE based on the experimental plasma density. Examples of XUV spectra are shown in Figs. 24,25. The individual harmonics are spectrally broadened and can be strongly distorted in shape, while this distortion fluctuates strongly from shot to shot. Furthermore, the spectral modulation in the XUV spectrum disappears for positively chirped incident pulses. Fluctuations of intensity and pulse duration are within 3% each and can be excluded as source of the fluctuating XUV spectra.

Chapter 6: Applications of Multi-TW, Few-cycle Light Pulses

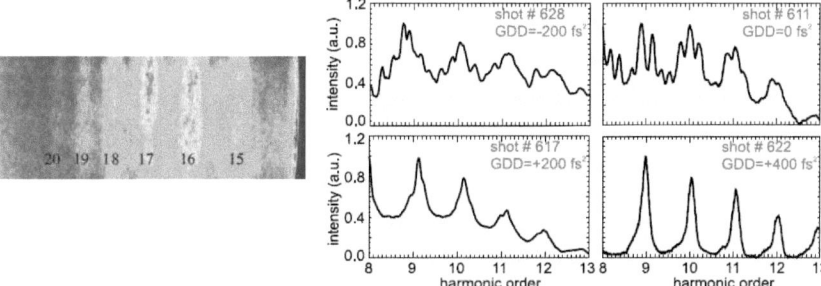

Fig. 25: Left: Harmonic spectrum up to the 20th harmonic corresponding to the highest harmonic expected from the CWE mechanism recorded in cooperation with R. Hörlein. Right: XUV spectra recorded for applied GDD via the AOM to the compressed 8 fs pulse of LWS-20 (taken from Ref. [B10]).

According to the 3-step model and PIC simulations, these observations can be a direct consequence of the few-cycle nature of the LWS-20 and its varying CEP [B10], which changes the relative phase between carrier wave and envelope of the incident light pulse. For this reason, the unequal temporal spacing between the electron bunches in the plasma, and correspondingly the event of XUV emission, is pronounced using few-cycle pulses and fluctuates with the varying CEP. Therefore, intrinsic properties of the CWE process could become apparent due to the few-cycle nature of the multi-TW light pulse of the LWS-20 NOPCPA system. However, the experimental proof of this hypothesis remains for future research: a locking and defined adjusting of the driver's CEP.

Already with this laser system, high-harmonics beyond the 20th harmonic could be observed, which are generated through the *Relativistic Oscillating Mirror* (ROM) mechanism [108], which is supposed to be superior with respect to CWE: pronounced reproducibility, higher photon energy, increased efficiency, smaller divergence and better phase locking properties. Further experimental investigations are on its way. However, SHHG still needs to fulfill its promise of higher efficiency compared to gas HHG.

A route to enhance the emission of ROM is a further energy-upscaling of the LWS NOPCPA system while even enhancing the temporal pulse contrast and solid-state targets with preshaped surface structure or liquid targets.

Chapter 7: Principles of Transient Absorption Spectroscopy

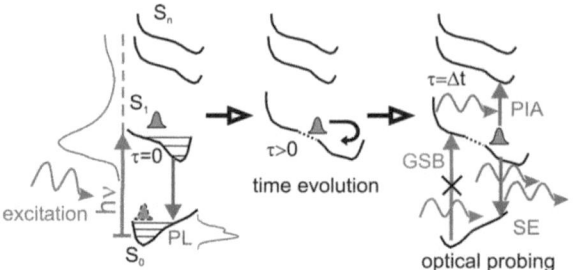

Fig. 26: Principle of transient absorption spectroscopy: photoexcitation, time evolution of wavepackets in excited states, optical probing at certain delay Δt, ground state bleach (GSB), photoinduced absorption (PIA), stimulated emission (SE).

The underlying principle of transient absorption (TA) spectroscopy is excitation of a molecular system and subsequently delayed probing of the photoexcited transient species. In this approach, the kinetics (formation and decay) of electronic states upon light absorption can be quantitatively determined within the experimental limits such as time resolution, maximum pump-probe delay and spectral probe detection bandwidth.

The Beer-Lambert law is a linear relationship between change in optical density ΔOD and photoinduced concentration change Δc of the corresponding transient species generated by photoexcitation with its inherent molar extinction coefficient ε_M

$$\Delta OD(t) = \varepsilon_M \cdot \Delta c(t) \cdot l = -\log(T(t)/T_0), \quad (7.1)$$

where l is the sample thickness. The transient change in optical density ΔOD is linked to the transmission T after photoexcitation and T_0 without excitation. T and T_0 are recorded during the TA measurements for various delays yielding a 2D TA map as shown in Fig. 27.

Fig. 26 points out the possible transient signals. Light absorption leads to a population of an excited state (S_n) and depletion of the ground state (S_0). For this reason, the excited system shows more probe transmission compared to the unexcited version yielding a negative transient signal ΔOD<0 termed *ground state bleach* (GSB) at the ground state absorption.

In the spectral range of *photoluminescence* (PL), the probe can induce a radiative recombination of the first ecited state to the ground state termed *stimulated emission* (SE) also yielding a negative transient signal ΔOD<0.

Newly occupied states following photoexcitation can possess transitions to higher-lying states associated to *photoinduced absorption* (PIA) with positive transient signals ΔOD>0. The initial photoexcited state can exhibit a transition to an energetically higher lying state, rendering an *excited state absorption* (ESA). Apart from that, it can undergo transitions to successive photoinduced species such as charged species (e.g., singlet-exciton dissociation to form free charges) or triplet states (e.g., singlet-exciton fission to yield two triplets or

Chapter 7: Principles of Transient Absorption Spectroscopy

intersystem crossing as a result of spin-orbit coupling). This ensemble of potential transient signals at different energetic positions constitutes the need of broadband probe detection to gain preferentially comprehensive knowledge of the studied molecular system.

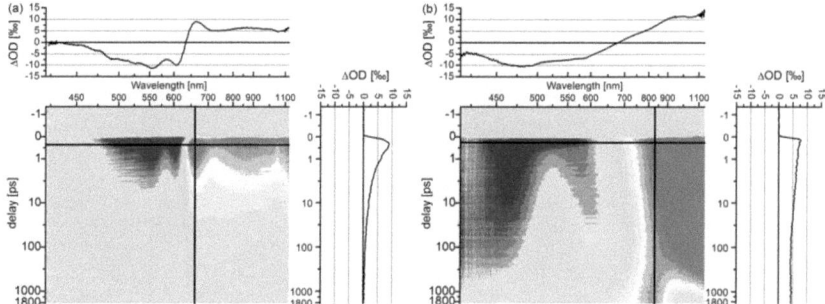

Fig. 27: Gap-free 2D TA map and lineouts for the TA measurement of (a) P3HT/Si film and (b) P3HT in chloroform solution.

Prior to retrieving the time evolution of the transient signals via reasonable fits, a detailed analysis of the TA spectra should be performed.

First, the steady-state absorbance and PL data of the actual investigated system should be compared to the TA signals to identify GSB and SE. Furthermore, absorption signals of charged species obtained in steady-state measurements are of high value to identify the corresponding transient photoinduced charged species.

Second, it needs to be verified if the spectral bands shift with increasing delay. There are a couple of origins for such a behavior. A spectral shift and narrowing is caused by vibrational cooling, where vibrational energy is dissipated to other normal modes and to the closeby environment. A spectral red-shift of bands such as SE can occur due to a conical intersection (CI) or due to the solvation process, where relaxation of polar solvent molecules after a change in charge distribution takes place in an excited solute. Moreover, migration of energy within species or energy transfer between compounds often leads to a spectral red-shift of the GSB within the corresponding characteristic time. Anisotropy can help to reveal the migration of photoexcitations.

Third, in case of samples that are a combination of different moieties, the TA spectra of the individual parts should be compared with the combination at various delays under the same experimental conditions. In this way, transient species (e.g., charge transfer states, cations, anions) assigned to the interaction between the moieties can readily be identified.

Fourth, in case of TA measurements of solid thin films also TA measurements of the same system - or a part of it – in solution can be of useful help to identify transient species in the film through the exclusion of various photoinduced processes of the film in the liquid but raising others at the same time. In general, the data evaluation should always be considered as system-dependent and therefore dynamic.

Chapter 8: UV-Vis-NIR TA spectrometer with 40 fs time resolution

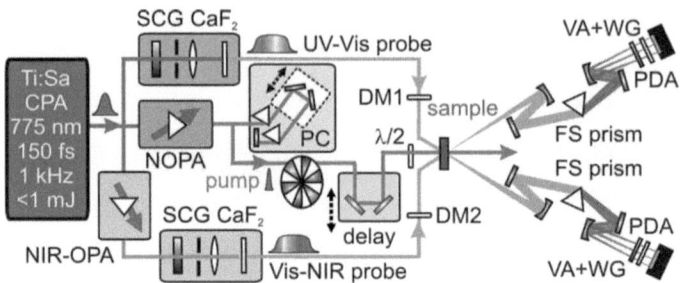

Fig. 28: Experimental layout of the ultrafast UV-Vis-NIR pump-probe spectrometer. CPA: chirped-pulse amplifier, NOPA: noncollinear optical parametric amplifier, SCG: supercontinuum generation, PC: prism compressor, DM: dichroic mirror, FS: fused silica, VA: variable attenuator, WG: wire-grid polarizer, PDA: photodiode array.

To investigate the nature of the photoexcitations in thin film photovoltaics and their inherent kinetics from fs up to several ns after visible excitation we developed a novel ultrafast pump-probe spectrometer. The gained understanding within the first part of my thesis work – i.e. generation of tunable ultrashort light pulses via NOPA/NOPCPA as well as supercontinuum generation – was a prerequisite for this development. The setup is described in Fig. 28. It allows for ultrafast ultrabroadband (UV-Vis-NIR) transient absorption (TA) experiments at room temperature. The time resolution was verified to be 40 fs, which is significantly better than for TA measurements in solution due to reduced group-velocity mismatch between pump and probe in the 100 nm thin film. The UV-Vis part of the setup employing a Ti:Sa chirped-pulse amplifier system (CPA 2001, ClarkMXR) with 1 kHz repetition rate as fundamental light source for pump and probe generation has been described in Ref. [38]. It was extended in the course of this work to allow the investigation of solid thin film samples and to expand the detection range into the NIR (Fig. 28, Fig. 29).

The details of this novel ultrafast broadband TA spectrometer are described in the corresponding publication in Appendix B7:

Role of Structural Order and Excess Energy on Ultrafast Free Charge Generation in Hybrid Polythiophene/Si Photovoltaics Probed in Real Time by Near-Infrared Broadband Transient Absorption
Daniel Herrmann, Sabrina Niesar, Christina Scharsich, Anna Köhler, Martin Stutzmann, and Eberhard Riedle
Journal of the American Chemical Society 133, 18220 (2011).

Briefly, a single-stage noncollinear optical parametric amplifier (NOPA) was used as pump source providing ultrashort pump pulses of 15 fs FWHM duration while the pump wavelength was chosen as 450, 518, 535, 555, 600 and 720 nm for site-selective excitation of the samples. To prevent photodegradation and to mimic the solar exposure of 1 kW/m^2 average power, low-excitation energies in the range of 4-60 nJ (4-60μJ/cm^2 excitation fluence) were used for P3HT/Si, while higher values (>300 nJ) were needed for DIP:C60 films due to its reduced absorbance. We routinely verified that the samples did not degrade due to the prolonged illumination by the pump pulses and only observed signal degradation when storing the samples over several months at ambient conditions in the lab.

Pump central wavelength	450-750 nm
Pump duration (FWHM)	15 fs
UV-Vis probe	280-740 nm
Vis-NIR probe	420-1150 nm
Spectral resolution	<3 nm
Repetition rate	1 kHz
Time resolution	40 fs
Detection sensitivity	ΔOD~10^{-4}
Pump diameter (1/e^2)	210 μm
Probe diameter (1/e^2)	110 μm

Tab. 1: Experimental parameters of the ultrafast pump, the broadband probe and the pump-probe measurements.

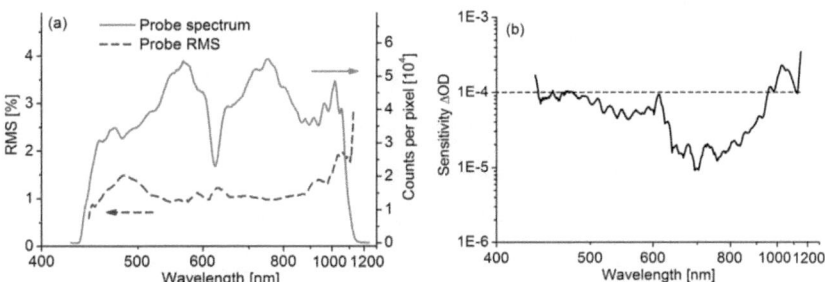

Fig. 29: (a) Broadband Vis-NIR probe spectrum and RMS noise (b) Experimental sensitivity of the Vis-NIR TA spectrometer calculated by taking the transient signal standard deviation per pixel at negative delays.

For gap-free TA measurements from 420 to 1150 nm (Fig. 29(a)), a novel probe setup was developed employing a cascade of supercontinuum generation (SCG) in YAG, OPA at 1180 nm in BBO and SCG in CaF$_2$ (Fig. 28). The broadband multichannel detection of the

TA signals employing a fused silica (FS) prism polychromator and a silicon photodiode array (PDA) based camera with 512 pixels is described in Ref. [38]. Typically, a probe RMS noise below 2%, a pump RMS noise of 1.5% and a probe detection sensitivity of $\Delta OD \sim 10^{-4}$ over the entire spectral range were achieved (Fig. 29). The wavelength calibration was performed with various filters and HR mirrors leading to a spectral resolution at the detector of better than 3 nm/pixel. To minimize pump straylight originating from the solid film samples, a broadband wire-grid polarizer (WG) was placed in front of the detector. This WG was aligned with maximum transmission for the probe polarization and the pump polarization was adjusted to be perpendicular to the probe polarization in this case. The residual straylight was deleted via subtracting the average signal over negative delays for each pixel, which also offers a tool to check for proper delay stage alignment.

If one assumes a Poisson distribution for the number of detected photons, the sensitivity is limited by:

$$|\Delta OD|_{min} = \left| -\log\left(\frac{I^*}{I_0}\right) \right| = \frac{\ln\left(1 + \frac{\Delta I}{I_0}\right)}{\ln(10)} \approx \frac{\Delta I}{I_0} \frac{1}{\ln(10)} = \frac{\sqrt{N}}{N} \frac{1}{\ln(10)} = \frac{1}{\sqrt{n \cdot m} \cdot \ln(10)},$$

where the minimum detectable change in photon number $\Delta I = I^* - I_0$ has to overcome the the standard deviation $\sqrt{N}$ of the Poisson distribution. $N = n \cdot m$, where n is the detected shot photon number per pixel and m is the number of averages. In the case of Fig. 29(a) with n=40000 and m=2000 the calculated sensitivity results in

$$|\Delta OD|_{min} = \left(\sqrt{40000 \cdot 2000} \cdot \ln(10)\right)^{-1} = 4,9 \cdot 10^{-5},$$

which is well within the experimentally determined sensitivity in Fig. 29(b).

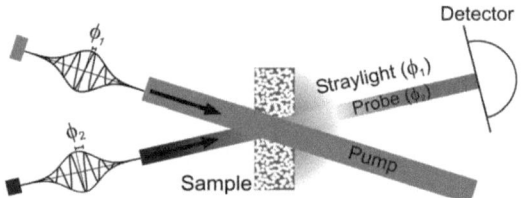

Fig. 30: A fraction of the pump straylight behind the sample is propagating along with the probe beam and leads to interferences at the detector. This shows that even the straylight still has the same defined phase relationship with the probe as the pump has, despite the cascaded optical nonlinearities and the stray processes in the sample.

Apart from providing the required high sensitivity for these experiments with low excitation fluence, an accurate calibration of the time zero is needed in order to resolve the primary photoexcitations in photovoltaics. For this reason, the raw data was corrected for the chirp of the probe pulses by adjusting the zero time at half rise of GSB and singlet-exciton signals or at coherent artifacts which occur due to two-photon absorption or wave-mixing [109]. Subsequently, a polynomial fit for the data points was applied. Additionally, we could show that the UV-Vis and the Vis-NIR pump-probe setup are interferometrically stable and thus offer the alternative to employ a periodic modulation due to spectral interference between pump straylight and probe at the detector to extract the delay via Fourier-transformation (Fig. 31). We have explicitly compared both options and found that the deviation of their result is negligible and enable calibration of the zero delay with a few-fs accuracy (Fig. 31). Overall, the time resolution of the ultrabroadband TA measurements was 40 fs, in particular because we do not need to measure over the spectral region of the SCG fundamental where a time jump of about 400 fs is present in the broadband probe pulse.

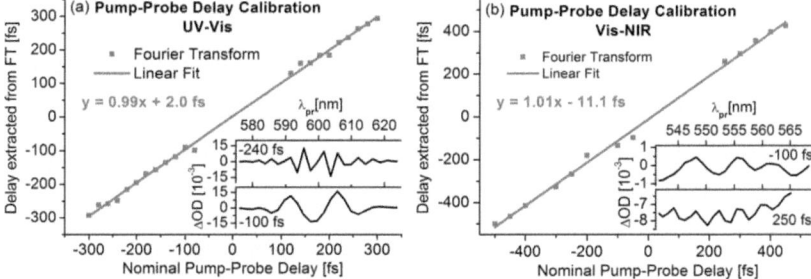

Fig. 31: Correlation between the nominal pump-probe delay and the delay retrieved from the interference fringes (insets) in the region of pump straylight on the detector employing Fourier transformation for the (a) UV-Vis and the (b) Vis-NIR TA spectrometer.

Chapter 9: Charge Generation and Recombination in Hybrid P3HT/Si Photovoltaics

9.1) Solar Cell Operation

To judge the prospects of photovoltaic materials for future energy supply, the achievable electrical output power vs. incident solar power has to be regarded. W. Shockley and H. Queisser performed an analysis termed *detailed balance limit of efficiency* of a p-n junction solar cell [110]. They assume, that the sun and the solar cell are black bodies with temperature T_s and T_c and that each absorbed photon with energy greater than the optical band gap energy $E_g = h\nu_g$ generates one electronic charge e at a certain open-circuit voltage V_{OC}.

First, they estimate the *ultimate efficiency u* through calculating the electrical output power devided by the incident light power. They assume a perfectly absorbing photovoltaic device yielding a voltage V_{OC} equal to the band gap potential $V_g = h\nu_g/e$ while employing the Planck distribution for the number of quanta at a certain frequency incident per unit area per unit time originating from a sourrounding black body sphere. The maximum efficiency is about 44%, which comes for a solar temperature of 6000 K and an optical band gap of 1.1 eV.

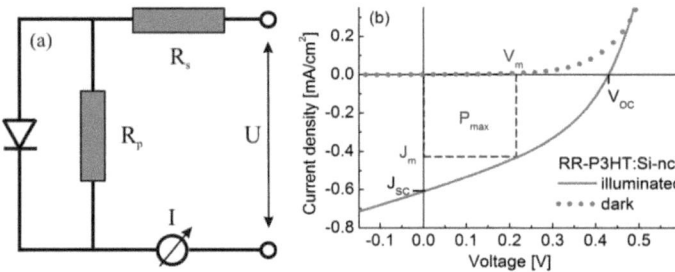

Fig. 32: (a) Equivalent circuit of a solar cell: diode, shunt resistance R_p, serial resistance R_s, current I and voltage U. (b) Current density (J) – voltage (V) curve of a P3HT/Si solar cell. Open-circuit voltage V_{OC}, short-circuit current density J_{SC}, current density (J_m) and voltage (V_m) at the maximum power point P_{max}.

Second, they calculate the *nominal efficiency u·ϕ·n*, which takes into account geometric considerations (such as small solid angle radiation from the sun), imperfect light absorption, and radiative as well as nonradiative processes. Fig. 32 shows the equivalent circuit for a solar cell and a typical current voltage curve. The removal from holes from the p-type region and electrons from the n-type region yields a current I. In this case, the voltage V_{OC} is a fraction n of V_g and approaches V_g at zero device temperature. The efficiency limit is then additionally bounded via deviding the only theoretically possible electrical power $V_{OC} \cdot I_{SC}$ by the incident solar power (Fig. 32):

Chapter 9: Charge Generation and Recombination in Hybrid P3HT/Si Photovoltaics

$$\frac{I_{SC} \cdot V_{OC}}{P_{in}} = u(T_s, E_g) \cdot \phi \cdot n(f, T_c, T_s, E_g), \qquad (9.1)$$

where I_{SC} is the photocurrent due to illumination (at V=0, short-circuit current), V_{OC} is the open-circuit voltage at I=0, ϕ is the probability - averaged over incident solar photons with photon energy above the band gap - that an incident photon will generate an electron-hole pair. f is a factor lumping together geometric factors, absorption probability ϕ and recombination processes [110].

Third, the actual maximum electrical output power for a realistic solar cell is derived by calculating the maximum rectangular below the current-voltage curve in Fig. 32(b) located at voltage V_m and current I_m. This yields a *fill factor FF* (or impedance matching factor) which describes the efficiency of the removal of holes from the p-type region and electrons from the n-type region:

$$FF = \frac{P_{max}}{I_{SC} \cdot V_{OC}} = \frac{I(V_m) \cdot V_m}{I_{SC} \cdot V_{OC}}. \qquad (9.2)$$

The actual maximum power conversion efficiency termed *detailed balance limit of efficiency* η thus results in:

$$\eta = \frac{P_{out}}{P_{in}} = \frac{I(V_m) \cdot V_m}{P_{in}} = u(T_s, E_g) \cdot \phi \cdot n(f, T_c, T_s, E_g) \cdot FF(f, T_c, T_s, E_g) = FF \cdot \frac{I_{SC} \cdot V_{OC}}{P_{in}}. \qquad (9.3)$$

The power conversion efficiency is therefore proportional to FF, I_{SC} and V_{OC}. An ideal silicon colar cell with E_g=1.1 eV, T_s=6000 K, T_c=300 K, normal incidence, perfect device absorption and radiative recombination as only recombination process for electron-hole pairs yields the maximum efficiency of η=30%, which is often quoted as *Shockley-Queisser-Limit* [110].

In Fig. 32(a), the shunt resistance R_p accounts for all nonradiative losses while the series resistance R_s represents bulk, contact and circuit resistivity. Therefore, the corresponding current equation according to Fig. 32(a) is:

$$I = I_{SC} + I_0 \left[\exp\left(\frac{e(U - R_s I)}{nkT}\right) - 1 \right] + \frac{U - R_s I}{R_p} \qquad (9.4)$$

where k is Boltzmann's constant, T is temperature, I_0 is the reverse-bias saturation current and $1 \leq n \leq 2$ is the diode ideality factor [111]. $R_p \to \infty$ and R_s=0 is taken for an ideal solar cell. A small R_p leads to a reduction of V_{OC} and the FF whereas an increase in R_s yields a decrease in I_{SC} and a dramatic reduction in FF by tilting the whole I-U curve around the point at V_{OC}.

Additional factors come into play in organic photovoltaics, which reduce the theoretical efficiency limit: disorder, low dielectric constant leading to a considerable exciton binding energy and losses during charge transport (nongeminate recombination etc.) yielding a reduced FF. Based on these circumstances particular for organic semiconductors, power conversion efficiencies of about 10-12% seem to be achievable under optimized solar cell conditions.

The experimental results of the following sections of chapter 9 are described in detail in the corresponding publication in Appendix B7:

Role of Structural Order and Excess Energy on Ultrafast Free Charge Generation in Hybrid Polythiophene/Si Photovoltaics Probed in Real Time by Near-Infrared Broadband Transient Absorption

Daniel Herrmann, Sabrina Niesar, Christina Scharsich, Anna Köhler, Martin Stutzmann, and Eberhard Riedle
Journal of the American Chemical Society 133, 18220 (2011).

9.2) Choice of Materials for Hybrid Photovoltaics

In this work we study the nature of the primary photoexcitations and their inherent kinetics in *hybrid P3HT/Si* thin film heterojunctions for photovoltaics by using ultrafast broadband transient absorption spectroscopy. The key factors determining the efficiency of free charge generation are identified.

The novel material combination P3HT/Si offers the advantage of combining the "best of both worlds" [112]. The semicrystalline polymer P3HT exhibits promising characteristics for optoelectronic applications such as high carrier mobilities, chemical stability and broad sun light absorption from the UV to the Vis [45,46,113,114]. Silicon adds to the widely used P3HT its own qualities such as abundant availability in various geometries, being non-toxic and a low band gap of 1.1 eV (indirect transition) leading to sun light absorption from the UV (direct band gap transitions) to the NIR (Fig. 34). Both materials can be processed from solutions for use in low-cost photovoltaic devices as alternative to conventional crystalline silicon solar cells.

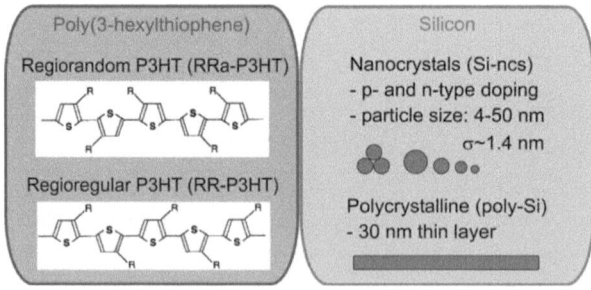

Fig. 33: Hybrid P3HT/Si heterojunctions: the unordered RRa-P3HT or the ordered RR-P3HT are combined with Si, either as few-nm small nanocrystals (Si-ncs) or as thin nm-scale polycrystalline layer (poly-Si).

Chapter 9: Charge Generation and Recombination in Hybrid P3HT/Si Photovoltaics

Due to sp^2-hybridization, *P3HT* builds π-bonds allowing for visible excitation. P3HT can form two distinct morphological phases associated with different chain conformations (Fig. 33). If the P3HT chain adopts a random coil conformation, the resulting film is amorphous. The associated absorption spectrum is unstructured and with a maximum centred around 450 nm (about 2.8 eV). This disordered structure prevails for regiorandom P3HT (RRa-P3HT, Fig. 2, Fig. 33), where steric repulsion between the hexyl side chains and the sulfur lead to a torsion of the polymer backbone [51,114], and is present in case of P3HT in diluted solution (Fig. 34). In regioregular P3HT (RR-P3HT, Fig. 2, Fig. 33), the polymer chains can planarize and assemble to form weakly coupled H-aggregates [115], which arrange in few-Å packed (Fig. 35(b)) two-dimensional lamellar structures via π-stacking [45,46,51,113,114]. As shown in Fig. 34, their spectroscopic signature is a red-shifted well-structured absorption spectrum with a 0-0 vibronic peak around 600 nm (about 2.0 eV) [115,B7], which is still a superposition of coil and aggregates.

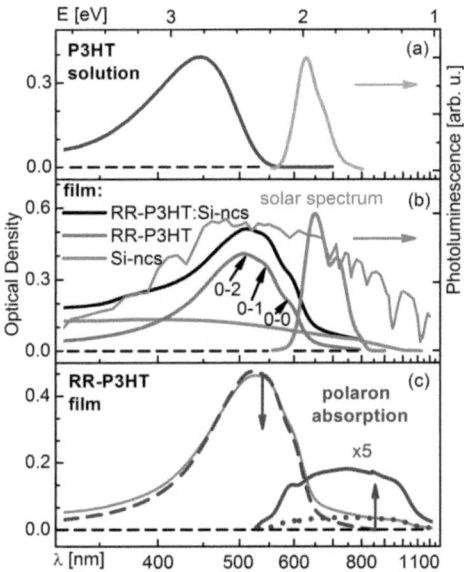

Fig. 34: Absorbance and photoluminescence spectra of P3HT in (a) dilute chloroform solution and (b) of RR-P3HT film. The absorbance of Si-ncs and RR-P3HT:Si-ncs (5:1) BHJ as film are also shown. (c) Absorbance spectra of RR-P3HT (blue dashes), oxidized RR-P3HT after adding FeCl3 (solid grey curve) and P3HT film polaron absorption (brown dots and brown solid curve).

In order to study the process of charge generation in these material systems, we not only need to know how many aggregated or coiled chains are present in the P3HT, but we also require a spectroscopic signature for charges in P3HT films. P3HT thin films can be chemically oxidized employing a strong oxidant [46]. Fig. 34(c) shows the absorbance spectrum of thus oxidized RR-P3HT thin films after dipping into solution of iron(III) chloride ($FeCl_3$) in acetonitrile (CH_3CN) followed by rinsing with acetonitrile to remove excess oxidant. From the raw spectrum of the treated film we obtain the P3HT polaron ($P3HT^+$) absorption as follows. The treatment with $FeCl_3$ decreases the known absorbance of the neutral P3HT molecule (blue dashes) and increases absorbance in the range from about 560 to 1150 nm (brown dots). The difference between the blue dashed line scaled to the peak of the grey solid curve in Fig. 34(c) and the grey solid curve reveals the RR-P3HT film polaron absorption (brown dots and brown solid line). It ranges from about 560 to 1150 nm with a characteristic shape and increases again up to the mid-infrared spectral region. The polaron absorption spectrum which we obtain in the P3HT films (Fig. 34(c)), closely matches the sub-gap polaron absorption bands obtained by cw photoinduced absorption (PIA) measurements as reported in previous publications [46,48,116]. Therefore the measured P3HT polaron absorption spectra can be compared with the TA spectra in Fig. 37.

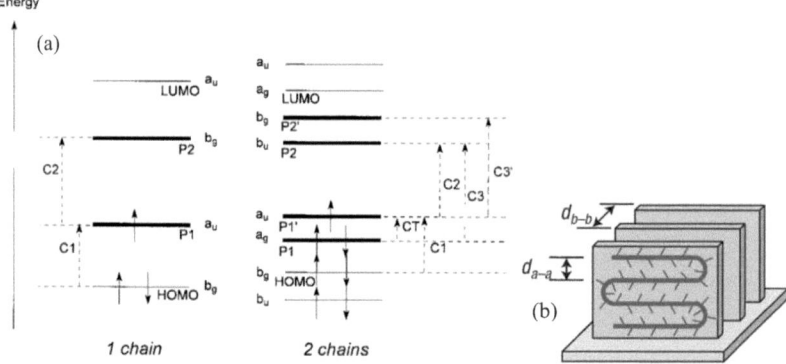

Fig. 35: (a) Scheme taken from Ref. [116] of the one-electron energy diagram for a polaron localized on a single conjugated chain (left), and delocalized over two cofacial chains (right). The aggregate possesses an almost filled band and an empty one with polaron absorption transitions C1 and C3,3' and the charge transfer transition CT. The symmetry of the orbitals and the relevant electronic excitations are indicated. (b) 2D lamellar arrangement of P3HT on the substrate, where the a-direction and b-direction are parallel and perpendicular to the thiophene ring plane, respectively (scheme taken from Ref. [51]. Ref. [114] states d_{a-a}=16.4 Å and d_{b-b}=3.8 Å for RR-P3HT).

RR-P3HT possesses comparable intra- and interchain excitonic coupling and charge transfer integrals allowing for light-induced energy and charge transfer between the planar chromophores [116a]. The term "*polaron*" is thereby applied to a charged polymer chain in a film, which interacts with its bulk surrounding via the polarizability to find its new energetic minimum. The polymer cation is the corresponding case of a single chain in solution, where the solvent interacts with the charged species. Extraction of one electron from a conjugated polymer chain induces two electronic levels $P_{1,2}$ in between the previously forbidden HOMO-LUMO band gap leading to two main sub-gap absorption bands in the mid-IR (C_1) and the Vis-NIR (C_2) spectral region as shown in Fig. 35 along with the corresponding optical transitions [46,116]. The polaron shows decreased bond-length alternations and typically extends over a few repeat units. Taking the increased interchain coupling of the ordered RR-P3HT lamellars (Fig. 35) into account leads to delocalization of molecular orbitals and a splitting of the electronic states yielding four polaronic levels equally shared by the neighboring chains with their new polaron transitions $C_{3,3'}$. The interchain interactions result in a weaker lattice relaxation in the aggregate, a polaron absorption red-shift by about 0.1-0.2 eV with an additional splitting and favoured 2D delocalized charge carriers. It was speculated that delocalization of the polaron over an infinite number of conjugated chains would lead to negligible relaxation energies, high band-like charge carrier moblities and low-energy charge-transfer excited state absorption evolving as Drude-like free charge carrier intraband absorption approaching the situation of conventional 3D inorganic semiconductors.

It was just recently shown that upon ordering P3HT, only the HOMO level shifts reducing the bandgap but not affecting the energy difference between P3HT LUMO and LUMO/CB of acceptor [117]. Fig. 36 shows that this leaves a range for the HOMO level of P3HT depending on the structural order. Furthermore, the energy difference of P3HT LUMO(D) and Si CB(A) of about 1.3 eV (neglecting additional influences on the band alignment) is a rather large driving force [57,118], facilitating efficient electron transfer from P3HT to Si.

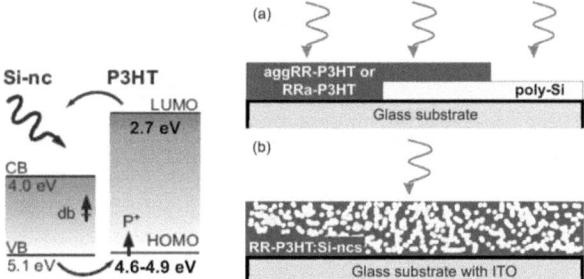

Fig. 36: Left: Proposed view of the energy levels below the vacuum level participating in charge transfer beween silicon and P3HT (scheme taken from Ref. [119] but values for P3HT are updated by recent results of Ref. [117]). Dangling bond defect (db) and polaron (P^+) are indicated. The scheme does not take into account potential additional influences on the band alignment such as Fermi level pinning or interfacial dipoles. Right: Sample architecture of the used hybrid P3HT/Si thin film (a) planar heterojunctions and (b) bulk heterojunctions on glass substrates [B7].

Additionally, several groups revealed a correlation between the open-circuit voltage V_{oc} and the HOMO(D)-LUMO/CB(A) energy difference [1,118,120], while Veldmann *et al.* (Δ =0.18 eV, [118]) and Scharber *et al.* (Δ =0.3 eV, [120]) found empirical formular for organic blends yielding different energy losses Δ:

$$eV_{oc} = |E_{HOMO}^{D} - E_{LUMO}^{A}| - \Delta. \tag{9.5}$$

Additionally, it is currently debated that a minimum LUMO(D)-LUMO(A) offset of 0.1-0.5 eV is needed as driving force for efficient electron transfer [118, 121]. These losses with respect to the band gap energy particulary in organic photovoltaics leads on the other hand to a similar situation as in a conventional p-n junction solar cell as described in section 9.1, where the open-circuit voltage is always lower than the bandgap potential V_g. In organic and inorganic solar cells V_{oc} is further lowered due to shunt losses (section 9.1).

A current-voltage curve of a RR-P3HT:Si-ncs solar cell is shown in Fig. 32(b). An open-circuit voltage V_{oc} of up to 0.75 eV for RR-P3HT:Si-ncs BHJs is measured [50], while this value is probably its inherent limit regarding Fig. 36 and currently more than achieved in polymer:PCBM blends [44,40]. Unfortunately, the HOMO(P3HT)-VB(Si) is rather small compared to many pure organic blends and can be detrimental for efficient hole transfer potentially leading to a reduced current in P3HT/Si solar cells. Furthermore, dangling bond defects know for Si-ncs can lead to enhanced recombination and ask for proper post-treatment such as etching and annealing of the Si-ncs after production in a microwave reactor. However, power conversion efficiencies of around 1% are achieved with preliminary devices [49].

The high electron affinity and delocalization as well as the screening that an electron experiences when being transferred has made PCBM the standard electron acceptor in organic photovoltaics. However, for investigations of the exciton dissociation mechanism, PCBM has the disadvantage of showing various TA signals from the Vis to the NIR which superimpose on the absorption by the exciton and polaron of P3HT (compare Fig. 37 and Fig. 41). The convolution of these signals renders it difficult to obtain a clear picture of the fundamental processes. *Silicon* – like PCBM - allows for a rapid electron delocalization after transfer which prevents back transfer and enables fast transport away from the interface. Yet, due to the higher dielectric constant of silicon, this effect should be even more pronounced. In contrast to PCBM, silicon does not show any photoinduced absorption signals in the spectral range of interest, so that the photophysics of charge generation can be studied clearly and unambiguously.

In our study (Fig. 33, Tab. 2), we employ 4-50 nm small Si-nanocrystals (Si-ncs, [122]) and 30 nm thin polycrystalline silicon (poly-Si, [123]) layers as the electron acceptor in order to study both film geometries of interest, *bulk heterojunctions* (BHJs) as well as *planar heterojunctions* (PHJs, [43]). For the development of efficient commercial solar cells, the BHJ structure is favoured since it offers a particularly large donor-acceptor interface. For studies that focus on the fundamental principles of the photophysics, the PHJ is advantageous for an unambigious data interpretation. To address the dependence of charge carrier generation on polymer structural order, disordered RRa-P3HT and semicrystalline RR-P3HT were used (Fig. 33, Tab. 2), while we also additionally enhanced the aggregation in RR-P3HT (aggRR-P3HT) by adding small amounts of poor solvent [B7,124].

Apart from the distinct optical effects of coiled and aggregated P3HT discussed above and shown in Ref. [B7], the structure of the films was carefully monitored via atomic force microscopy (AFM). The corresponding topographical images in 2D and 3D plots are shown in Ref. [B7] and reveal an increased aggregation from RRa- to aggRR-P3HT.

Sample	Abbreviation
RR-P3HT (BASF), spun from CHCl$_3$	RR-P3HT
RRa-P3HT (Bayreuth), spun from CHCl$_3$	RRa-P3HT
RR-P3HT (Rieke), spun from CHCl$_3$:EtAc	aggRR-P3HT
silicon nanocrystals	Si-ncs
polycrystalline silicon	poly-Si
RR-P3HT:Si-ncs bulk heterojunction	RR-P3HT:Si-ncs BHJ
RRa-P3HT/poly-Si planar heterojunction	RRa-P3HT/poly-Si PHJ
aggRR-P3HT/poly-Si planar heterojunction	aggRR-P3HT/poly-Si PHJ

Tab. 2: List of samples used in this work along with their abbreviations.

9.3) Probing the Ultrafast Photovoltaic Charge Generation Mechanism in Real-time

The thin film samples were designed to enable direct comparison between pure P3HT, pure Si and the P3HT/Si heterojunction in ultrafast TA spectroscopy at the same experimental conditions by moving the relevant sample regions into the pump-probe region. We choose pump and probe beam diameters at the sample that allow for ensemble averaging over the finely grained morphology of the thin films to mimic their use as photovoltaic device and to ensure low excitation fluence (9 µJ/cm^2) comparable to the solar exposure.

Fig. 37(a) shows typical Vis-NIR TA spectra at 300 fs, 20 ps and 1-2 ns pump-probe delays here for aggP3HT/poly-Si PHJ (red solid curve) compared with the TA spectra of the corresponding pure polymer film (black solid curve). The excitation was 9 µJ/cm^2 at 518 nm. For each measurement, the pairwise difference between TA spectra of the heterojunction and the pure polymer is calculated (green solid curve). For comparison, the P3HT film polaron absorption spectrum (blue dashes) is included. In order to directly compare the transient spectra of pure P3HT with the corresponding P3HT/Si heterojunction (and therefore account for small variations in experimental conditions between the measurements), the latter is scaled to yield the same numer of initial photoexcitations at the earliest time (0-80 fs). The integral over the GSB area equals the product of the excitation density times the strength of the first electronic transition of P3HT regardless of small variations of Franck-Condon activity due to differing morphologies of P3HT in the various samples. The GSB integral at the earliest time possible therefore represents a good relative measure of the number of initial photoexcitations. This analysis allows to reveal signal changes solely inherent to the different nature of the samples.

In the TA spectra of P3HT/Si thin films without an applied external field, we can identify the transient signatures of P3HT polarons (620-900 nm), P3HT singlet-excitons (900-1150 nm) and P3HT GSB (420-625 nm). No stimulated emission and no triplet formation is observed, which we both detected for P3HT in solution [B7].

Adding Si to ordered and unordered P3HT leads to more than a factor 2 increased polaron yield compared to the pure polymer. This is visualized by the green curve in Fig. 37(a), where Si simultaneously enhances the polaron signal and reduces the singlet-exciton absorption throughout the whole investigated time range up to 2 ns.

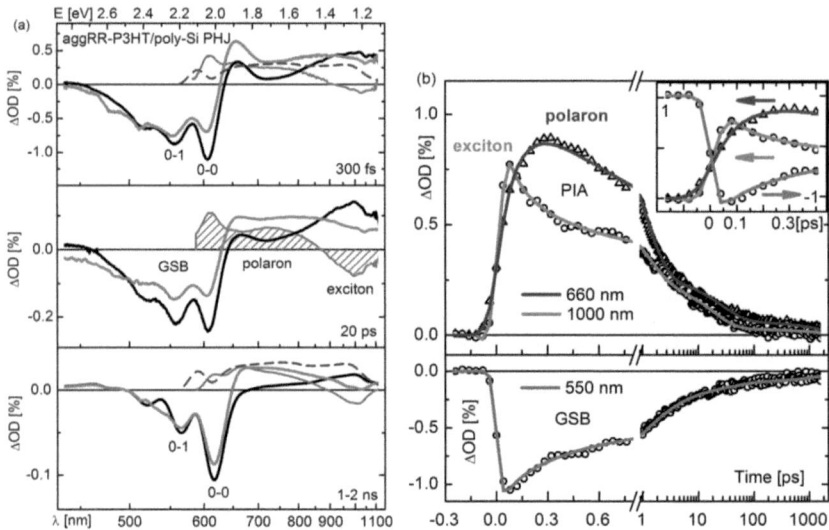

Fig. 37: (a) TA spectra of aggRR-P3HT/poly-Si PHJ for 300 fs, 20 ps and 1-2 ns pump-probe delay. The spectra are scaled to the same number of initial photoexcitations. The scaled P3HT film polaron absorption spectrum (blue dashes) and the difference (green solid curve) between pure P3HT (black solid curve) and P3HT/Si (red solid curve) are compared. (b) Experimental TA signals (open symbols) as function of the pump-probe delay with corresponding fits (solid curves) of GSB (green), singlet-exciton (red) and polaron (blue) for aggRR-P3HT/poly-Si PHJ.

As we are particularly interested in the initial signal changes, we treat the ultrafast component separately by fitting the transient signals according to the function

$$\Delta OD = A_1 \cdot \exp(-t/\tau_1) + A_{SE} \cdot \exp(-t/\tau_{SE})^\beta + const. \tag{9.6}$$

while Fig. 37(b) shows typical kinetics of the P3HT transient species along with the fit curves based on the parameters listed in Tab. 3.

Chapter 9: Charge Generation and Recombination in Hybrid P3HT/Si Photovoltaics

The GSB shows the distinct vibronic structure of the P3HT aggregates and occurs instantaneously, i.e. within the time resolution of our pump-probe setup of 40 fs. The singlet-exciton shows very similar ultrafast kinetics as the GSB and is formed within 40 fs. Subsequently, both transient signals decrease with an initial rapid exponential decay with a time constant of 140 fs followed by a stretched exponential decay with a mean decay constant of $<\tau>=1.3$ ps.

The polaron signature shows distinctly different dynamics. The signal shows a delayed rise with a time constant of 140 fs and reaching its maximum at 300 fs. The observed time constant of 140 fs therefore clearly correlates the delayed formation of polarons due to the initial decay of singlet-excitons.

Parameter	Polaron (660 nm)	Exciton (1000 nm)	GSB (550 nm)
A_1	$-14.6 \cdot 10^{-3}$	$2.6 \cdot 10^{-3}$	$-1.6 \cdot 10^{-3}$
τ_1	0.14 ps	0.14 ps	0.14 ps
A_{SE}	$20.2 \cdot 10^{-3}$	$7.1 \cdot 10^{-3}$	$-13.6 \cdot 10^{-3}$
τ_{SE}	0.41 ps	0.63 ps	0.6 ps

Tab. 3: Fit parameters: amplitudes and time constants for the transient species in Fig. 37.

This charge generation is followed by a stretched exponential decay with a mean decay constant of $<\tau>=0.8$ ps. Importantly, there is no significant polaron formation present directly after photoexcitation, i.e. at 40 fs. The polaron yield is about a factor 2 higher in hybrid P3HT/Si using aggregated RR-P3HT compared to employing unordered RRa-P3HT which is dominated by coils.

This ultrafast GSB recovery can be readily understood by an additional ultrafast nonradiative decay mechanism of at least some of the excited P3HT molecules. A minimal rate model (see appendix A3) that allows relaxation of the exciton into the polaron by dissociation and to the P3HT ground state by nonradiative decay directly renders the result that the yield of each channel is given by the ratio of the individual rate to the sum of both. The spectroscopic signal is additionally weighted by the respective extinction coefficients. The exciton serves as a reservoir and the same femtosecond kinetics is observed for the decay of the exciton signal and the recovery of the GSB. Nonradiative electronic decay on the femtosecond time scale is now widely reported for a large variety of molecular systems and believed to be frequently mediated by conical intersections [4]. Whether a conical intersection is also responsible for the observed ultrafast nonradiative decay in P3HT films has to be clarified in the future. It has recently been established in conjugated polymers that exciton localization occurs in tens of femtoseconds and leads to nonemissive states [2,B7].

The main decrease of the transient signals on the longer timescale can be fitted by a stretched exponential due to an ensemble of recombination times as a result of an ensemble of P3HT chain conjugation lengths and on-chain vs. interchain recombination processes [125].

The constant accounts for long-lived species such as long-lived charges that can readily be extracted at electrodes.

These experiments probe the ultrafast charge generation mechanism in P3HT and P3HT/Si for the first time and in real-time revealing instantaneous formation of singlet-excitons in P3HT as primary photoexcitation which dissociate to form polarons in 140 fs. The speed of the electron transfer from P3HT to Si implies a strong exchange integral of the excited state orbitals between donor and acceptor.

9.3.1) Quantitative Analysis

Absorbed photons	$2.6 \cdot 10^9$ / pulse
Areal density of excitations	$2.4 \cdot 10^{13}$ cm^{-2}
Initial charge density	$6 \cdot 10^{18}$ cm^{-3}
Polaron cross section	$(3.4 \pm 2) \cdot 10^{-16}$ cm^2
Polaron extinction coeff.	$(4 \pm 1) \cdot 10^4$ M^{-1}cm^{-1}

Tab. 4: Parameters of the P3HT/Si heterojunction with 9 µJ/cm^2 excitation at 518 nm.

For an absorbed number of photons of $2.6 \cdot 10^9$ per pulse at an excitation of 9 µJ/cm^2 at 518 nm, we obtain an excited state areal density of $2.4 \cdot 10^{13}$ cm^{-2} (Tab. 4,[B7]). The supporting information of Ref. [B7] contains the derivation of the polaron molar extinction coefficient ε_M of $(4 \pm 1) \cdot 10^4$ M^{-1} cm^{-1}, the polaron cross section σ of $(3.4 \pm 2) \cdot 10^{-16}$ cm^2, and the initial charge density of $6 \cdot 10^{18}$ cm^{-3} from our TA measurements. These values are similar to previous investigations of RR-P3HT and RR-P3HT:PCBM blends [54,59,126].

The analysis of the TA measurements revealed formation of singlet-excitons as primary photoexcitation upon light absorption. As shown in Fig. 38, these bound electron-hole pairs can dissociate into positive (P3HT$^+$) and negative (P3HT$^-$) P3HT polarons with quantum yield δ and the remaining photoexcitations can decay to the ground state with yield 1- δ through release of vibrational energy (P3HT + Energy).

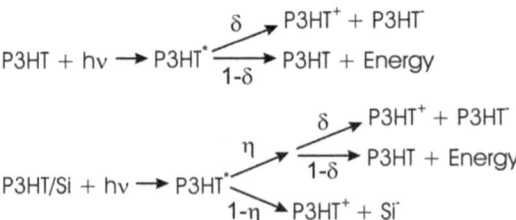

Fig. 38: Yields of transient species generated by photoexcitation of pure P3HT and P3HT/Si heterojuntions with P3HT as electron donor and Si as electron acceptor.

Chapter 9: Charge Generation and Recombination in Hybrid P3HT/Si Photovoltaics

In the P3HT/Si heterojunction, a proportion η of the photons is absorbed so far away from the hybrid heterojunction as to lead to the same result as in pure P3HT. The remaining fraction (1- η) of photons is absorbed near a P3HT/Si interface undergoing ultrafast electron transfer to the Si with 140 fs forward transfer time. It is possible to estimate the initial yield of polarons by considering the relative magnitudes of the TA signals at 300 fs given in Fig. 37(a) [58]. To determine the quantum yields, we first consider the TA signal magnitude at 300 fs due to singlet-excitons in aggRR-P3HT/poly-Si PHJ, which typically has about 0.75 times the magnitude as for pure aggRR-P3HT. Therefore, the fraction of photoexcitations that initially do not sense the Si component is taken as η=0.75. The maximal magnitude of the TA signal due to polarons at 300 fs is typically about 2.2 times that for pure P3HT, which results in δ=0.17.

While we obtain a quantum yield of 17% for the formation of P3HT$^+$ polarons in a pure film of aggRR-P3HT, this value raises to 38% in the aggRR-P3HT/poly-Si PHJ in combination with a Si$^-$ anion yield of 25% (Tab. 5). We note that in the heterojunction device, the P3HT layer covering the silicon has a film thickness of only 40 nm, so that excitons are created close to the donor-acceptor interface. The significant enhancement of exciton dissociation through the addition of silicon implies that silicon performs well as an electron accepting material in agreement with the discussion of Fig. 37. The yields and time constants point to efficient electron transfer and a spatial singlet-exciton delocalization in ordered P3HT of about 10 nm in agreement with previous investigations [55,127].

Sample	P3HT$^+$	P3HT$^-$	P3HT+E	Si$^-$
aggP3HT	0.17	0.17	0.83	0
aggP3HT/Si	0.38	0.13	0.62	0.25

Tab. 5: Initial quantum yields of charges in pure P3HT and aggRR-P3HT/poly-Si PHJ.

9.4) Role of Polymer Structural Order and Excess Energy

Site-selective excitation of P3HT/Si reveals more than a factor of 2 enhancement of initial polaron yield by only increasing the pump wavelength from 450 to 600 nm under the otherwise identical experimental conditions (Fig. 39). Furthermore, the present TA experiments verify that excitation above 550 nm selectively excites P3HT aggregates while excitation below that increasingly excites P3HT coils [B7]. In particular, exciting RR-P3HT at 600 nm leads to a complete lack of coil contributions to the GSB of TA spectra from 60 fs on [B7]. The GSB of RR-P3HT excited at 450 nm initially shows significant coil excitation, while these contributions slowly vanish within 13 ps. From that time on, the bandwidth of the

GSB is independent of the excitation wavelength revealing an energy transfer from P3HT coils to aggregates with an 1/e transfer time of 3 ps. Moreover, the present TA measurements with varied excitation fluence identify bimolecular nongeminate recombination - pointing to free charges - in P3HT/Si with aggregated polymer (RR- and aggRR-P3HT/Si) and monomolecular geminate recombination - revealing bound polaron pairs - for the coiled (RRa-P3HT/Si) version. These findings point to a significant role of polymer structural role rather than vibrational excess energy on the efficiency of free charge generation. We record the same factor of 2 more efficient charge generation in hybrid P3HT/Si heterojunctions employing the aggregated polymer compared to using the unaggregated version by two different and independent methods: (i) by using different P3HT configurations and therefore solely changing the structural order in the P3HT film [B7], and (ii) by solely changing the excitation wavelength and therefore selectively exciting defined P3HT regions (Fig. 39). If vibrational excess energy was required for the exciton dissociation process as previously suggested for other systems [52,55-57], we would expect a high initial polaron yield for excitation at 450 nm and a lower polaron yield for 600 nm pump wavelength. Since this is not the case, we can conclude that the inherent driving force for electron transfer in P3HT/Si heterojunctions is sufficient.

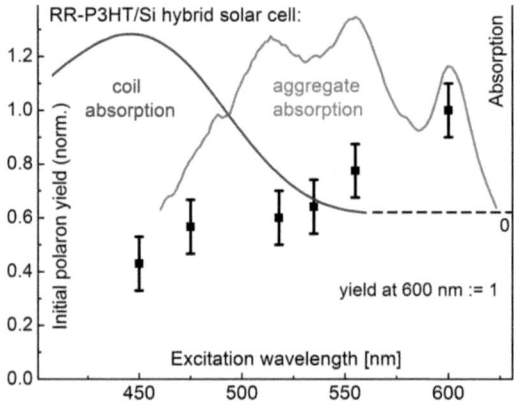

Fig. 39: Dependence of initial polaron yield (black squares) on excitation wavelength for hybrid RR-P3HT/Si heterojunctions. The initial polaron yield was normalized to 1 at 600 nm excitation. Selective excitation of coiled (P3HT in diluted solution, blue curve) vs. aggregated (RR-P3HT film GSB with 600 nm excitation, red curve) RR-P3HT domains reveals greater than a factor of 2 more efficient charge separation if exciting directly the aggregated RR-P3HT domains.

Chapter 10: Effect of Morphology in DIP/C$_{60}$ Photovoltaics

As detailed in section 9.1, a crucial limit of any single band gap photovoltaic device operating under solar exposure is the lack of harvesting the entire solar spectrum and the full energy of supra-bandgap photons. A possible way to circumvent this limitation is to generate multiple electron-hole pairs from an absorbed high-energy photon. Multiple generation of excitons has been shown for inorganic semiconductor nanocrystals but so far not for organic semiconductors [128]. However, singlet-exciton fission to form two triplet excitons is an available alternative in conjugated organics to enhance the quantum yield of charges by up to a factor of 2, if the energy of the singlet-exciton is at least twice the energy of the lowest lying triplet state. In this case it is energetically favourable and spin-allowed for the singlet to undergo fission and form two triplets [129]. This has been shown for acene molecules even recently in photovoltaic systems [130,131]. For the material diindenoperylene (DIP) in combination with the fullerene C$_{60}$ this has also been proposed [W. Brütting, private commun. 2010], since in this case the so far unknown DIP triplet energy would be in resonance with the C$_{60}$ LUMO eventually promoting efficient electron transfer from DIP to the fullerene. Apart from this specific mechanism, the more conventional process of dissociation of singlet-excitons to form charges in DIP/C$_{60}$ is also possible. To resolve this issue, the PIA of DIP singlet- and triplet-excitons as well as the DIP cation should be known.

In any case, first DIP/C$_{60}$ planar (PHJ) and planar-mixed heterojunctions (PMHJ) have shown to be rather promising solar cells as presented in Fig. 40 and Ref. [132] and DIP exciton diffusion lengths of about 90 nm were measured [private commun. 2011]. The U-J curves indicate a strong dependence of J_{sc} and FF on the heterojunction morphology. First TA measurements on the route to reveal the primary photoinduced processes in DIP/C$_{60}$, with the aim to correlate them to the U-J curves, are presented here.

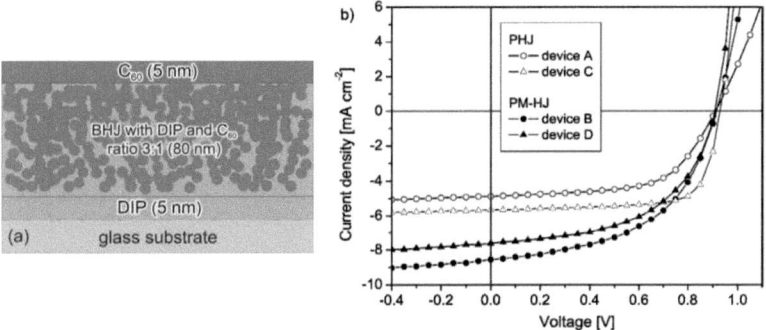

Fig. 40: (a) Schematic sample architecture of a DIP:C60 PMHJ used in this work. (b) Current-voltage curves taken from Ref. [132] for DIP/C60 PHJ and DIP:C60 PMHJ showing different Jsc and FF. Thickness: A/B 65 nm, B/C 130 nm

In this work, thin films of pure DIP, pure C$_{60}$ (not PCBM), DIP/C$_{60}$ PHJ and DIP:C$_{60}$ PMHJ are probed by the ultrafast Vis-NIR pump-probe spectrometer presented in chapter 8. The results of the individual samples are shown in Fig. 41 along with the steady-state absorbance (OD) and PL. At first glance it becomes evident that this material combination shows transient signals over the entire probe spectral range and even exceeding it at both edges. Fig. 41(a) shows that the GSB of DIP shows the distinct vibronic signatures, which are superimposed by a TA signature from 450-600 nm. DIP also possesses a broad PIA in the NIR, which can be fitted by a stretched exponential with a mean decay time of about 2.5 ps. The pure GSB signal below 440 nm shows a similar decay time. Both transient signals contain a delayed signal change (additional decay for NIR PIA and rise for GSB), which can be fitted by an exponential with a ca. 40 ps time constant. Fig. 41 shows that the sum (black dotted curve) of the TA spectra of the individual pure DIP and pure C_{60} film does not match the TA spectra of the heterojunctions (black solid curve) in the spectral region from 450-600 nm. Furthermore, the fraction of photoexcitations leading to the NIR PIA almost entirely vanishes within 1 ns. The PHJ and PMHJ contain the same amount of DIP and C_{60} as the pure films but exhibit different interface architecture of electron donor (DIP) and acceptor (C_{60}).

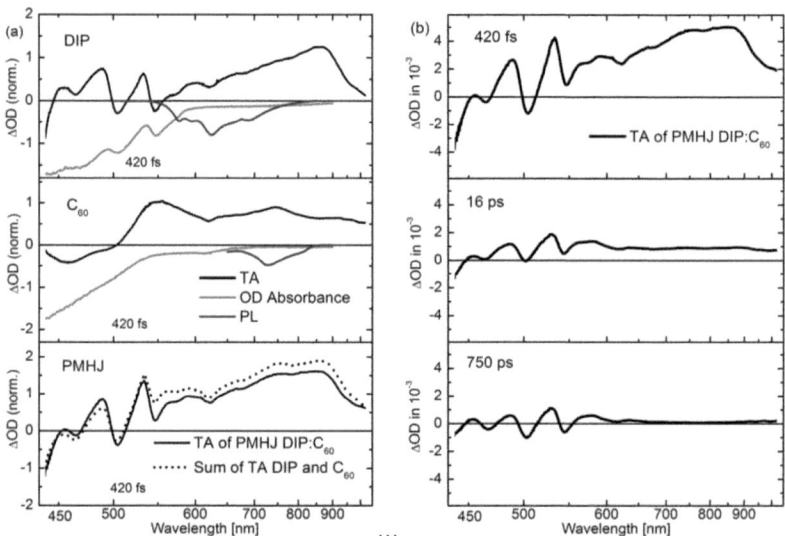

Fig. 41: (a) TA spectra of films of pure DIP (black, top) and pure C_{60} (black, middle) and the PMHJ (black solid curve, bottom) all at 420 fs. Black dotted curve: sum of the TA spectra of pure DIP and pure C_{60} at the same conditions. Steady-state absorbance (OD, red) and PL (blue) as comparison. (b) The raw TA spectrum of DIP:C60 PMHJ for 420 fs, 16 ps and 750 ps. Excitation of 760 nJ at 550 nm.

The absorption band from 450-600 nm, which is superimposed on the DIP/C_{60} GSB is revealed by subtracting an according to the steady state absorbance of DIP/C_{60} scaled version of the pure GSB signal at 440 nm from this composed signature. Fig. **42**(a) shows the resulting PIA signal of DIP, PHJ and PMHJ for various excitation energies. It indicates that adding electron acceptor C_{60} to DIP in a PHJ, the initial PIA signal magnitude can be enhanced. A further enhancement can be observed by changing the film morphology from PHJ to PMHJ yielding a factor of 2 higher signal magnitude compared to pure DIP.

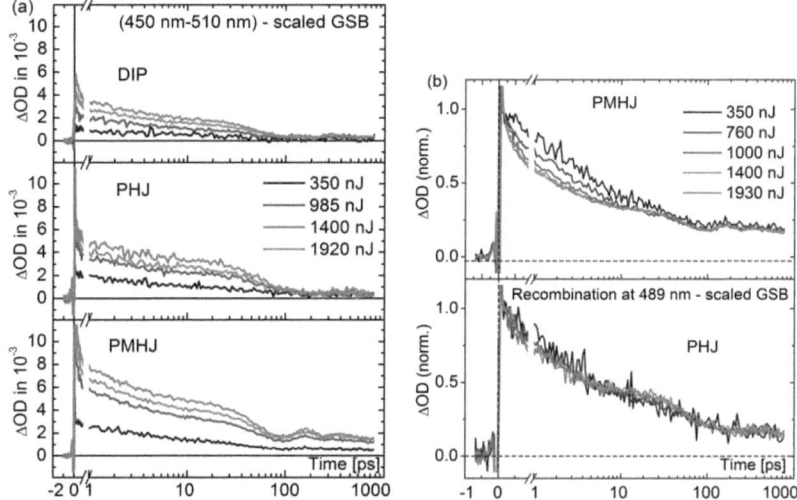

Fig. 42: (a) TA signal between 450 and 510 nm minus the scaled GSB (possibly "cation" yield) for pure DIP, DIP/C60 PHJ and DIP:C60 PMHJ for various excitation energies (b) Effect of morphology on the decay rate of the normalized "cation"-signal for the PHJ and the PMHJ as function of varied excitation energies.

After normalizing these temporal PIA traces against their initial signal magnitudes reveals a pronounced bimolecular nongeminate recombination for the PMHJ compared to the PHJ. The decay of this PIA signal superimposed on the GSB can be fitted by a stretched exponential with a mean decay time constant of 7 ps and a delayed monoexponential rise with a significant amplitude and an about 10 ps time constant. If one compares these results with the black dotted curve in Fig. 41(a), one concludes an ultrafast and a delayed formation of the PIA superimposed on the GSB. Summing up, this PIA from 450-600 nm superimposing the GSB is likely to be a transient signature assigned to charges (possibly the DIP cation since the removal of only one electron should result in a PIA near the ground state absorption of neutral DIP, however the DIP film remains to be chemically oxidized). Its behaviour upon adding C_{60} and as function of varied film morphology matches well the results obtained in Fig. 40 for the U-I performance. In this context, it could be concluded that the addition of C_{60} leads to an electron transfer from DIP to the fullerene, which is more efficient (by 50%) in the PMHJ as in the PHJ due to the increased donor-acceptor interface leading to increased J_{sc}. This in turn leads to pronounced bimolecular recombination in the PMHJ and therefore to a reduced FF as also becomes visible in Fig. 40(b) and Fig. 42(b).

Chapter 11: Conclusions and Outlook

In the course of this dissertation the nature of some of the fastest light-matter interactions was studied. For this goal the development of spectrally tunable few-cycle light sources was advanced approaching two optical cycles in the Vis and NIR spectral range.
Few-cycle light pulses were generated in a broad parameter range from nJ to 100 mJ pulse energies and from 100 kHz to 10 Hz repetition rates. Diagnostics to characterize few-cycle light pulses in the time and frequency domain were developed to gain insight into phase effects. Fundamental knowledge and limits of noncollinear optical parametric chirped-pulse amplification (NOPCPA) was attained through numerical simulations and experimental investigations. The pursued studies led to the currently most intense few-cycle light pulse worldwide (7.9 fs, 130 mJ, 16 TW, 805 nm) with high temporal pulse contrast approaching the limits of NOPCPA. The technique of NOPCPA was shown to be scalable in energy up to the 200 mJ range. These few-cycle light pulses were used to generate high-harmonic radiation from solid surfaces in the range of 40 to 100 nm in the regime of coherent wake emission (CWE) and even approaching the relativistic oscillating mirror (ROM) mechanism. The generation of multi-TW, few-cycle light pulses allowed to reveal the sub-cycle nature of the harmonic emission and consequently opens the route to intense single attosecond pulses for XUV-pump/XUV-probe spectroscopy. Additionally, these light pulses were employed to accelerate electrons from underdense plasmas yielding 10 fs, 30 MeV, few-pC electron bunches for use in ultrafast electron diffraction or free electron lasers (FEL) [133,134].
To broaden the gain bandwidth of NOPCPA even in the few-cycle limit, novel schemes employing two pump beams at the same pump wavelength in one parametric amplifier and two pump wavelengths in two cascaded NOPCPA stages were characterized. It was found that composed signal spectra via both approaches possess a slowly varying spectral phase and could therefore be compressed. With the first approach the generation of mJ-level, 7 fs light pulses is achieved. The second approach was found to be a route for generating octave-spanning, sub-5 fs pulses on the multi-mJ level. In this context, the existence of a parametric phase imprinted on the signal during amplification so as to compensate for wavevector-mismatch is proven. These approaches are alternatives to complex coherent wavelength multiplexing using fiber lasers or OPCPA achieving single-cycle waveforem synthesis [8,9]. A theoretical analysis of effective phase-mismatch including this parametric phase is developed to predict the gain bandwidth in NOPCPA. It is shown that the obtained predictions match the experimental signal bandwidths very well. Moreover, a strong pump wavelength-dependence of the parametric phase is found and its challenging consequences for multi-stage NOPCPA chains reaching saturation and CEP-control are outlined.

I propose to set up a NOPCPA laser system starting with a Yb-based fiber or thin disc laser as common light source providing 50 µJ pulses at 1035 nm showing a few-100 fs pulse duration at a few-kHz repetition rate. About 0.5 µJ can be used to generate a supercontinuum (SC: 460-1020 nm) from bulk such as YAG or sapphire crystals. The main part of the remaining pump can be frequency-doubled yielding about 20 µJ at the SH for pumping of a type-I BBO NOPA. Seed and pump pulse durations can be matched by using a prism precompressor for the SC. This approach can provide a high-contrast seed for a multi-stage NOPCPA chain containing about 2 µJ energy, which was shown to be compressable to sub-10 fs light pulses even at higher energies [86] but needs to be (negatively) stretched for efficient NOPCPA. The 2-4 cascaded NOPCPA stages should be operated at decreasing moderate ($\leq 10^3$) gain but with a saturated last amplifier offering enough gain to dermine the spectral and spatial signal properties [B5]. A fraction of the fundamental pump light can be used to synchronize a high-energy Yb-based diode-pumped CPA laser for pumping the NOPCPA chain with few-ps pulses containing J-level energies. Its pulse duration should be a compromise between providing high temporal signal pulse contrast (as well as high-gain broadband parametric amplification) and minimized temporal jitter usually of about 100-200 fs. A hybrid bulk/dispersive mirror compressor can be used for final pulse compression.

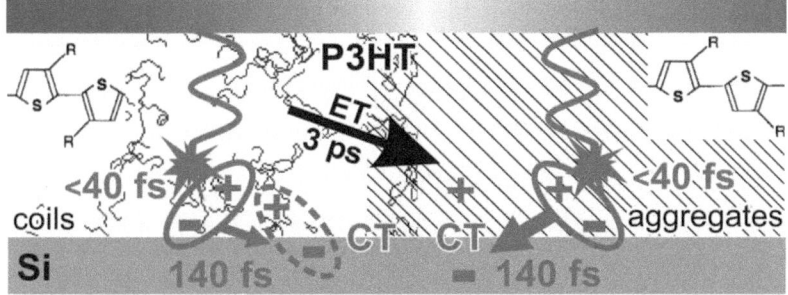

Fig. 43: Primary photoinduced processes in hybrid P3HT/Si thin film heterojunctions.

The gathered knowledge of NOPCPA in combination with supercontinuum generation spanning two octaves from the UV to the NIR paved the way to probe ultrafast primary photoinduced processes in organic materials for photovoltaic applications as shown in Fig. 43 An ultrabroadband (UV-Vis-NIR) pump-probe spectrometer with a high time resolution of 40 fs was realized for these experiments. In the TA experiments, we can identify the transient signatures of P3HT polarons (630-900 nm) and singlet-excitons (900-1150 nm). Ultrafast TA of the photoinduced species in hybrid Polythiophene/Silicon (P3HT/Si) films revealed the formation of singlet-excitons as primary, instantaneous photoexcitation in the organic semiconductor P3HT, which dissociate to form polarons with a time constant of 140 fs. The addition of Si to RR-P3HT or RRa-P3HT significantly enhances the polaron yield in the active layer as a consequence of electron transfer, which reveals its excellent suitability as electron acceptor. Silicon is a crystalline semiconductor allowing for highly efficient charge delocalization. It is available abundantly and has a higher dielectric constant as compared to the state-of-the-art electron acceptor PCBM thus allowing for even improved screening of the electron preventing back transfer. The speed of the charge transfer (CT) from P3HT to Si indicates a strong exchange integral of the excited state orbitals between donor and acceptor and implies the prospect of high efficiencies in corresponding photovoltaic devices. However, the formation of charge carriers in organic solar cells is found to be *delayed* and could resolve a long lasting debate about the primary photoexcitation in organic semiconductors.

The real-time probing of the charge generation mechanism is a result of the high time resolution and the very broad probe bandwidth of the developed pump-probe arrangement employing parametric amplifiers and supercontinua. Spectrally tunable, 15 fs light pulses allowed to reveal the crucial inherent parameters yielding efficient free charge generation in P3HT/Si, which is required for efficient solar cells as future solution for sustainable energy supply. Hybrid sub-100 nm thin films with various sample architecture (planar vs. bulk heterojunction) were probed with varied P3HT structural order (RRa-, RR- and aggregated RR-P3HT). An energy transfer (ET) from coiled to aggregated P3HT with a time constant of 3 ps is resolved. Polymer structural order and therefore the mobility of electrons *and* holes to enable a high initial electron-hole separation is found to be the key criterion for efficient free charge generation and suppressed geminate recombination of bound polaron pairs as potential loss mechanism. Quantitatively, it is found that aggregated P3HT leads to about a factor 2 higher polaron quantum yield compared to using coiled P3HT in P3HT/Si and shows bimolecular recombination as main loss mechanism in solar cells. Additional supply of vibrational excess energy is found not to be a crucial necessity for efficient electron transfer from P3HT to Si and therefore indicates that the inherent driving force is already sufficient. For this reason, pure organic and hybrid photovoltaic devices using P3HT should employ highly aggregated P3HT. The loss of the coils as light-harvesting complex and their high-energy absorption can be compensated by stacking heterojunctions in tandem or even multiple solar cells using conjugated polymers with different band gaps [135]. Ongoing RR-P3HT/Si-ncs device optimization led to 1% power conversion efficiency but further optimization of the Si-ncs properties and the BHJ interpenetrating network is required.

First ultrafast broadband TA experiments were performed on promising organic DIP/C_{60} photovoltaics. Transient signatures of charges were obtained. It is found that their yield and recombination mechanism strongly depend on the heterojunction morphology. Adding C_{60} to DIP in a planar-mixed heterojunction enhances the yield of this species by a factor of 2, while in turn a pronounced bimolecular nongeminate recombination is identified. These findings on the microscopic scale are in good match with the macroscopic current-voltage performance of the corresponding photovoltaic devices.

I propose the following experiments to further advance the understanding of photoconversion in hybrid and organic photovoltaics:

P3HT/Si:
- TA of P3HT polarons in the ns to µs range to determine the lifetime of charge carriers in P3HT/Si. Vary the temperature of the film to judge if long-lived species are trapped and can be thermally activated.

- BHJ with sub-10 nm Si-ncs of varied phosphor doping density and varied post-processing (HF etching, annealing) to reduce surface and bulk defects.
- Vary P3HT:Si-ncs mixing ratio to judge if a lower Si-ncs content increases the polaron formation time via pronounced contributions of exciton diffusion.
- Nanostructures of varied size (10-50 nm) should be obtained in the P3HT film. Choose spin-coating of Si-ncs dissolved in ethanol or spray-coating to cover structured P3HT layer and achieve percolation paths throughout the BHJ [136].
- Ultraviolet Photoelectron Spectroscopy (UPS) is needed to determine the Si CB and VB and to judge if the bandalignment between P3HT and Si can be optimized.

Polymer/PCBM:
- Does the dissociation of singlet-excitons to form polarons also hold true for other organic solar cells? Investigate the primary photoconversion mechanism in various polymer:PCBM films. Interesting polymers: MeLPPP, PIF, DOO-PPD, PF2/6 and low-bandgap polymers such as PCBTBT. Does the polaron formation time depend on the LUMO(polymer)-LUMO(PCBM) energy gap; how much driving force is needed for efficient electron transfer?
- Vary polymer:PCBM mixing ratio to judge if a lower acceptor content increases the polaron formation time via pronounced contributions of exciton diffusion.
- Probe charge transfer excitons (CTE) in polymer:PCBM films.

Singlet-fission in oligomers:
- The absolute triplet energy level in polymers/oligomers is usually at around 1.5 eV following $T_1 = |S_1 - S_0|_{0-0}^{PL} - 0.7eV\,(\pm 0.1 eV)$ [köh09]. Oligomers could be investigated if they yield singlet-fission to form two triplet states and therefore enhance the charge quantum yield by up to a factor of 2.

DIP/C_{60}:
- UV-Vis TA measurements of DIP, C60, DIP:C_{60} in film and solution
- TA measurements of DIP and DIP:C_{60} films with varied excitation wavelength; in particular at 430 nm (possible CT state transition).
- Anisotropy measurements in film and solution.

TA measurements in reflection with applied electric field:
- Apply external electric field on electrodes of solar cells and measure TA in reflection (double pass).

Bibliography

[1] Brabec, C. J.; Zerza, G.; Cerullo, G.; De Silvestri, S.; Luzzati, S.; Hummelen, J. C.; Sariciftci, S. Chem. Phys. Lett. **2001**, 340, 232-236.

[2] Collini, E.; G. D. Scholes Science **2009**, 323, 369-373.

[3] Krausz, F.; Ivanov, M.; Rev. Mod. Phys. **2009**, 81, 163.

[3a] Sundström, V.; Annu. Rev. Phys. Chem. **2008**, 59, 53.

[3b] Zewail, A. H.; J. Phys. Chem. A **2000**, 104, 5660.

[4] Polli, D. *et al.* Nature **2010**, 467, 440.

[5] Goulielmakis, E. *et al.* Science **2008**, 320, 1614.

[6] Baltuška, A.; and Kobayashi, T. Appl. Phys. B **2002**, 75, 427.

[7] Brida, D.; Marangoni, M.; Manzoni, C.; De Silvestri, S.; and Cerullo, G. Opt. Lett. **2008**, 33, 2901.

[8] Krauss, G.; Lohss, S.; Hanke, T.; Sell, A.; Eggert, S.; Huber, R.; and Leitenstorfer, A. Nature Photon. **2010**, 4, 33.

[9] Huang, S.-W. *et al.* Nature Photon. **2011**, 5, 475.

[10] Maiman, T. H. Nature **1960**, 187, 493.

[11] Köcher, W.: *Solid-State Laser Engineering*, Springer, Heidelberg 2006.

[12] Physik-Journal 9 (2010) Nr. 7: *Schwerpunkt 50 Jahre Laser*, Wiley-VCH 2010.

[13] Myer, J.A.; Johnson, C.L.; Kierstead, E.; Sharma, R.D.; and Itzkan, I. Appl. Phys. Lett. **1970**, 16, 3.

[14] Spence, D.E.; Kean, P.N.; and Sibbett, W. Opt. Lett. **1991**, 16, 4244.

[15] Rausch, S.; Binhammer, T; Harth, A.; Kärtner, F. X.; and Morgner U. Opt. Express **2008**, 16, 17411.

[16] Strickland, D.; and Mourou, G. Opt. Commun. **1985**, 56, 219.

[17] Perry, M. D. *et al.* Opt. Lett. **1999**, 24, 160.

[18] Kiriyama H. *et al.* Opt. Lett. **2010**, 35, 1497.

[19] Gaul, E. Appl. Opt. **2010**, 49,1676.

[20] Kiriyama H. *et al.* Opt. Lett. **2008**, 33, 645.

[21] Nisoli, M.; De Silvestri, S.; and Svelto, O. Appl. Phys. Lett. **1996**, 68, 2793.

[22] Hauri, C. P. *et al.* Appl. Phys. B **2004**, 79, 673.

[23] Cavalieri, A. L. *et al.* N. J. Phys. **2007**, 9(7), 242.

[24] Park, J.; Lee, J. H.; and Nam, C. H. Opt. Lett. **2009**, 34, 2342.

[25] Gale, G.M.; Cavallari, M.; Driscoll, T.J.; and Hache, F. Opt. Lett. **1995**, 20, 1562.

[26] Wilhelm T.; Piel J.; and Riedle E. Opt. Lett. **1997**, 22, 1494.

[27] Shirakawa, A. and Kobayashia, T. Appl. Phys. Lett. **1998**, 72, 147.

[28] Cerullo, G.; Nisoli, M.; Stagira, S.; and De Silvestri, S. Opt. Lett. **1998**, 23, 1283.

[29] Adachi, S. *et al.* Opt. Express **2008**, 16, 14341.
[30] Dubietis, A.; Jonusauskas, G.; and Piskarskas, A. Opt. Commun. **1992**, 88, 437.
[31] Ross, I. N.; Matousek, P.; New, G. H. C.; Osvay, K. J. Opt. Soc. Am. B **2002** 19, 2945.
[32] Dubietis, A.; Butkus, R.; and Piskarskas, A. IEEE J. of Sel. Top. in Quantum. Electron. **2006**, 12, 163.
[33] Ishii, N. *et al.* Opt. Lett. **2005**, 30, 567.
[34] Witte, S.; Zinkstok, R. Th.; Wolf, A. L.; Hogervorst, W.; Ubachs, W.; and Eikema, K. S. E. Opt. Express **2006**, 14, 8168.
[35] Tavella, F.; Nomura, Y.; Veisz, L.; Pervak V.; Marcinkevicius A.; and Krausz, F. Opt. Lett. **2007** 32, 227.
[36] Piskarskas, A.;_Butkus, R. CLEO/Europe 2011 paper CG4_1.
[37] Manzoni, C.; Polli, D.; and Cerullo G. Rev. Sci. Instr. **2006**, 77, 023103.
[38] Megerle, U.; Pugliesi, I.; Schriever, C.; Sailer, C. F.; Riedle, E. Appl. Phys. B **2009**, 96, 215.
[39] Thompson, B. C.; Fréchet, J. M. J. Angew. Chem. Int. Ed. **2008**, 47, 58.
[40] Park, S. H.; Roy, A.; Beaupré, S.; Cho, S.; Coates, N.; Moon, J. S.; Moses, D.; Leclerc, M.; Lee, K.; Heeger, A. J. Nature Photon. **2009**, 3, 297.
[41] Gur, I.;Fromer, N. A.; Geier, M. L.; Alivisatos, A. P. Science **2005**, 3, 462.
[42] Tang, C. W.; Appl. Phys. Lett. **1986**, 48, 183.
[43] Sariciftci, N. S.; Braun, D.; and Zhang, C.; Srdanov, V. I.; Heeger, A. J.; Stucky, G.; and Wudl, F. Appl. Phys. Lett. **1993**, 62, 585.
[44] Brabec, C. J.; Gowrisankar, S.; Halls, J. J. M.; Laird, D.; Jia, S.; Williams, S. P. Adv. Mater. **2010**, 22, 3839.
[45] Sirringhaus, H. *et al.* Nature **1999**, 401, 685.
[46] Österbacka, R.; An, C. P.; Jiang, X. M.; Vardeny, Z. V. Science **2000**, 287, 839.
[47] Huynh, W. U.; Dittmer, J. J.; Alivisatos A. P. Science **2002**, 295, 2425.
[48] Oosterhout, S. D.; Wienk, M. M.; van Bavel, S. S.; Thiedmann, R.; Koster, L. J. A.; Gilot, J.; Loos, J.; Schmidt, V.; Janssen, R. A. J. Nature Mater. **2009**, 8, 810.
[49] Liu, C.-Y.; Holman, Z. C.; Kortshagen, U. R. Nano. Lett. **2009**, 9, 449.
[50] Niesar, S.; Dietmueller, R.; Nesswetter, H.; Wiggers, H.; Stutzmann, M. Phys. Status Solidi A **2009**, 206, 2775.
[51] Kim, Y. *et al.* Nature Mater. **2006**, 5, 197.
[52] Clarke, T.; Ballantyne, A. M.; Nelson, J.; Bradley, D. D. C.; Durrant, J. R. Adv. Funct. Mater. **2008**, 18, 4029.
[53] Mauer, R.; Howard, I. A.; Laquai, F. J. Phys. Chem. Lett. **2010**, 1, 3500.
[54] Howard, I. A.; Mauer, R.; Meister, M.; Laquai, F. J. Am. Chem. Soc. **2010**, 132, 14866.
[55] Köhler, A.; dos Santos, D. A.; Beljonne, D.; Shuai, Z.; Brédas J,-L.; Kraus, A.; Müllen, K.; Friend, R. H. Nature **1998**, 392, 903.
[56] Arkhipov, V. I.; Emelianova, E. V.; Bässler, H. Phys. Rev. Lett. **1999**, 82, 1321.

[57] Ohkita, H. et al. J. Am. Chem. Soc. **2008**, 130, 3030.

[58] Piris, J.; Dykstra, T. E.; Bakulin, A. A.; van Loosdrecht, P. H. M.; Knulst, W.; Trinh, M. T.; Schins, J. M.; Siebbeles, L. D. A. J. Phys. Chem. C **2009**, 113, 14500.

[59] Guo, J.; Ohkita, H.; Benten, H.; Ito, S. J. Am. Chem. Soc. **2010**, 132, 6154.

[60] Banerji, N.; Cowan, S.; Leclerc, M.; Vauthey, E.; Heeger, A. J. J. Am. Chem. Soc. **2010**, 132, 17459.

[61] Etzold, F.; Howard, I. A.; Mauer, R.; Meister, M.; Kim, T.-D.; Lee, K.-S.; Baek, N. S.; Laquai, F. J. Am. Chem. Soc. **2011**, 133, 9469.

[62] Baumgartner, R. A.; and Byer, R. L. Optical Parametric Amplification, IEEE J. Quantum Electron. **1979**, 15, 432444.

[63] Riedle, E.; Beutter, M.; Lochbrunner, S.; Piel, J.; Schenkl, S.; Spörlein, S.; Zinth, W.; Appl. Phys. B **2000**, 71, 457.

[64] Reider, G.: *Photonik: eine Einführung in die Grundlagen*, Springer Wien 1997.

[65] Boyd, R. W.: *Nonlinear Optics*, Academic Press San Diego/London 2003.

[66] Renault, A.; Kandula, D. Z.; Witte, S.; Wolf, A. L.; Zinkstok, R. Th.; Hogervorst, W.; and Eikema, K. S. E. Opt. Lett. **2007**, 32, 2363.

[67] Baltuska, A.; Fuji, T.; and Kobayashi, T. Phys. Rev. Lett. **2002**, 88, 133901.

[68] Wang, J.; Dunn, M. H.; and Rae, C. F. Opt. Lett. **1997**, 22, 763.

[69] Oien, A. L.; McKinnie I. T.; Jain P.; Russell, N. A.; and Warrington, D. M. Gloster, L. A. W. Opt. Lett. **1997**, 22, 859.

[70] Rustad, G.; Arisholm, G.; and Farsund, Ø. Opt. Express 2011, 19, 2815.

[71] Liao, Z. M.; Jovanovic, I.; Ebbers, C. A.; Fei, Y.; Chat, B. Opt. Lett. **2006** 31, 1277.

[72] Arisholm, G. J. Opt. Soc. Am. B **1997**, 14, 25432549.

[73] Tavella, F.; Marcinkevicius, A.; and Krausz, F. New J. Phys. 2006, 8, 219.

[74] Witte, S. et al. Appl. Phys. B **2007**, 87, 677.

[75] Louisell, W. H., Yariv, A., and Siegman, A. E., Phys. Rev. **1961**, 124, 1646.

[76] Mollow B. R.; and Glauber R. J. Phys. Rev. **1967**, 160, 1076.

[77] Kleinman, D. A. Phys. Rev. **1968**, 174, 1027.

[78] Dorrer, C.; Begishev, I. A.; Okishev, A. V.; and Zuegel, J. D. Opt. Lett. **2007**, 32, 2143.

[79] Bagnoud, V.; Zuegel, J. D.; Forget, N.; and Le Blanc, C. Opt. Express **2007**, 15, 5504.

[80] Gu, X. et al. Opt. Express **2009**; 17, 62.

[81] Harris, S. E.; Oshman, M. K.; and Byer, R. L. Phys. Rev. Lett. **1967**, 18, 732.

[82] Carrion, L.; and Girardeau-Montaut, J.-P. J. Opt. Soc. Am. B **2000**, 17, 78.

[83] Picozzi, A.; and Haelterman, M. Phys. Rev. E **2001**, 63, 056611.

[84] Tavella, F.; Marcinkevicius, A.; and Krausz, F. Opt. Express **2006**, 14, 12822.

[85] Bradler, M.; Baum, P.; Riedle, E. Appl. Phys. B **2009**, 97, 574.

[86] Antipenkov R.; Varanavičius A.; Zaukevičius A.; and Piskarskas A. P. Opt. Express **2011**, 19, 3519.

[87] Wood, W.M.; Siders, C.W.; and Downer, M.C. Phys. Rev. Lett. **1991**, 67, **3523**.
[88] Trushin, S. A. *et al.* Appl. Phys. B **2005**, 80, 399.
[89] Ahmad, I.; Trushin, S.A.; Major, Z.; Wandt, C.; Klingebiel, S.; Wang, T.-J.; Pervak, V.; Popp, A.; Siebold, M.; Krausz, F.; Karsch, S. Appl. Phys. B **2007**, 97, 529.
[90] Teubner, U; Gibbon, P. Rev. Mod. Phys. **2009**, 81, 445.
[91] Kaluza, M.; Schreiber, J.; K. Santala, M. I.; Tsakiris, G. D.; Eidmann, K.; Meyer-ter-Vehn, J.; and Witte, K. J. Phys. Rev. Lett. **2004**, 93, 045003.
[92] Herrmann, D. Master Thesis, University of Texas at Austin, August **2007**.
[93] Kane, D. J. IEEE J. of Sel. Top. in Quantum Electron. **1998**, 4, 278.
[94] Forget N.; Cotel A.; Brambrink E.; Audebert P.; and Le Blanc C.; Jullien A.; Albert O.; and Chériaux G. Opt. Lett. **2005**, 30, 2921.
[95] Didenko N.V.; Konyashchenko A.V.; Lutsenko A.P.; Tenyakov, S.Yu. Opt. Express **2008**, 16, 3178.
[96] Meshulach, D.; Barad, Y.; and Silberberg, Y. J. Opt. Soc. Am. B **1997**, 14, 2122.
[97] Antonetti, A.; Blasco, F.; Chambaret, J. P.; Cheriveux, G.; Darpentigny, G.; Le Blanc, C.; Rousseau, P.; Ranc, S.; Rey, G. ; Salin, F. Appl. Phys. B **1997**, 65, 197.
[98] Tavella, F.; Schmid, K.; Ishii, N.; Marcinkevicius, A.; Veisz, L.; Krausz, F.; Appl. Phys. B **2005**, 81, 753.
[99] Jullien, A.; Kourtev, S.; Albert, O.; Cheriaux, G.; Etchepare, J.; Minkovski, N.; Saltiel, S.M. Appl. Phys. B **2006**, 84, 409.
[100] Tang, Y.; Ross, I. N.; Hernandez-Gomez, C.; New, G. H. C.; Musgrave, I.; Chekhlov, O. V.; Matousek, P.; and Collier, J. L. Opt. Lett. **2008**, 33, 2386.
[101] Sosnowski, T. S.; Stephens, P. B.; and Norris, T. B. Opt. Lett. **1996**, 21, 140.
[102] Zeromskis, E.; Dubietis, A.; Tamosauskas, G.; and Piskarskas, A. Opt. Commun. **2002**, 203, 435.
[103] Tajima, T.; Dawson, J. M. Phys. Rev. Lett. **1979**, 43, 267.
[104] Esarey, E.; Schroeder, C. B.; and Leemans, W. P. Rev. Mod. Phys. **2009**, 81, 1229
[105] Osterhoff, J. *et al.* Phys. Rev. Lett. **2008**, 101, 085002.
[106] Pukhov A.; Meyer-ter-Vehn, J. Appl. Phys. B **2002**, 74, 355.
[107] Nomura, Y. *et al.* Nature Phys. **2009**, 5, 124.
[108] Tsakiris, G. D.; Eidmann K.; Meyer-ter-Vehn J.; and Krausz F. New J. Phys. **2006**, 8, 19.
[109] Lorenc, M.; Ziolek, M.; Naskrecki, R.; Karolczak, J.; Kubicki, J.; Maciejewski, A. Appl. Phys. B **2002**. 74, 19.
[110] Schockley, W.; and Queisser, H. J. J. Appl. Phys. **1961**, 32, 510.
[111] Brabec, C.; Dyakonov, V.; Parisi, J.; Sariciftci, N. S.: *Organic Photovoltaics,* Springer, Berlin/Heidelberg 2003.
[112] Jabbour, G. E.; and Doderer, D. Nature Photon. **2010**, 4, 604.
[113] McCullough, R. D.; Tristram-Nagle, S.; Williams, S. P.; Lowe, R. D.; Jayaraman, M. J. Am. Chem. Soc. **1993**, 115, 4910.

[114] Chen, T.-A.; Wu, X; Rieke, R. D. J. Am. Chem. Soc. **1995**, 117, 233-244.

[115] Clark, J.; Silva, C.; Friend, R. H.; Spano, F. C Phys. Rev. Lett. **2007**, 98, 206406.

[116] Beljonne, D.; Cornil, J.; Sirringhaus, H.; Brown, P. J.; Shkunov, M.; Friend, R. H.; and Brédas J.-L. Adv. Funct. Mater. **2001**, 11, 229.

[116a] Köse, M. E.; J. Phys. Chem. C **2011**, 115, 13076.

[117] Tsoi, W. C. *et al.* Macromolecules **2011**, 44, 2944.

[118] Veldman, D., Meskers, S. C. J.; and Janssen R. A. J. Adv. Funct. Mater. **2009**, 19, 1939.

[119] Dietmueller, R. *et al.* Appl. Phys. Lett. **2009**, 94, 113301.

[120] Scharber, M. C.; Mühlbacher, D.; Koppe, M.; Denk, P.; Waldauf, C.; Heeger, A. J.; and Brabec, C. J. Adv. Mater. **2006**, 18, 789.

[121] Schueppel, R. *et al.* Phys. Rev. B **2008**, 77, 085311.

[122] Knipping, J.; Wiggers, H.; Rellinghaus, B.; Roth, P.; Konjhodzic, D.; Meier, C. J. Nanosci. Nanotechnol. **2004**, 4, 1039-1044.

[123] Scholz, M.; Gjukic, M.; Stutzmann, M. Appl. Phys. Lett. **2009**, 94, 012108.

[124] Campbell, A. R.; Hodgkiss, J. M.; Westenhoff, S.; Howard, I. A.; Marsh, R. A.; McNeill, C. R.; Friend, R. H.; Greenham, N. C. Nano Lett. **2008**, 8, 3942.

[125] Movaghar, B.; Grünewald, M.; Ries, B.; Bässler, H.; Würtz, D. Phys. Rev. B **1986**, 33, 5545.

[126] Shuttle, C. G.; O'Regan, B.; Ballantyne, A. M.; Nelson, J.; Bradley, D. D. C.; Durrant, J. R. Phys. Rev. B **2008**, 78, 113201.

[127] Dogariu, A.; Vacar, D.; Heeger, A. J. Phys. Rev. B **1998**, 58, 10218.

[128] Beard, M.C.; Knutsen, K.P.; Yu, P.; Luther, J.M.; Song, Q.; Metzger, W.K.; Ellingson, R.J.; Nozik, A.J. Nano Lett. **2007**, 7, 2506.

[129] Smith, M. B.; and Michl, J. Chem. Rev. **2010**, 110, 6891.

[130] Rao, A.; Wilson, M. W. B.; Hodgkiss, J. M.; Albert-Seifried, S.; Bässler, H.; and Friend, R. H. J. Am. Chem. Soc. **2010**, 132, 12698.

[131] Wilson, M. W.; Rao, A.; Clark, J.; Kumar, R. S. S.; Brida, D.; Cerullo, G.; Friend, R. H. J. Am. Chem. Soc. **2011**, 133, 11830.

[132] Wagner, J. *et al.* Adv. Funct. Mater. **2010**, 20, 4295.

[133] Srinivasan, R.; Feenstra, J. S.; Park, S. T.; Shoujun, X.; Zewail, A. H. Science **2005**, 307, 558.

[134] Fuchs, M. *et al.* Nature Physics **2009**, 5, 826.

[135] Gilot, J.; Wienk, M. M.; Janssen, R. A. J. Appl. Phys. Lett. **2007**, 90, 143512.

[136] Wiedemann, W. *et al.* Appl. Phys. Lett. **2010**, 96, 263109.

[137] Köhler, A.; Bässler, H. Mater. Sci. Engineering R **2009**, 66, 71.

[138] Dmitriev, V.G.; Gurzadyan, G.G.; and Nikogosyan, D.N.: *Handbook of nonlinear optical crystals*, Springer (2006).

[139] Liu, H. J.; Chen, G. F.; Zhao, W.; Wang, Y. S.; Wang, T.; Zhao, S. H. Opt. Commun. **2001**, 197, 507

Appendix

A1) Derivation of group velocity matching between idler and signal:

Parallel and perpendicular component of wavevector mismatch in Fig. 4:

$$\Delta k_\| = k_{p\|} - k_{s\|} - k_{i\|} = k_p - k_s \cos(\alpha) - k_i \cos(\beta(\omega)), \quad \Delta k_\perp = k_{s\perp} - k_{i\perp} = k_s \sin(\alpha) - k_i \sin(\beta(\omega))$$

I one assumes a monochromatic pump: $\dfrac{\partial k_p}{\partial \omega} = \dfrac{\partial \alpha}{\partial \omega} = 0$

In the following alle derivations against the angular frequency are evaluated at the central frequency ω_0.

The phase-matching condition: $\quad \Delta k_0 = 0 \wedge \dfrac{\partial \Delta k_\|}{\partial \omega} = \dfrac{\partial \Delta k_\perp}{\partial \omega} = 0$

$$\frac{\partial \Delta k_\|}{\partial \omega} = -\frac{\partial k_s}{\partial \omega}\cos(\alpha) + k_i \sin(\beta)\frac{\partial \beta}{\partial \omega} - \frac{\partial k_i}{\partial \omega}\cos(\beta) \overset{!}{=} 0$$

$$\frac{\partial \Delta k_\perp}{\partial \omega} = \frac{\partial k_s}{\partial \omega}\sin(\alpha) - k_i \sin(\beta)\frac{\partial \beta}{\partial \omega} - \frac{\partial k_i}{\partial \omega}\sin(\beta) \overset{!}{=} 0$$

Now, multiply the first equation by $\cos(\beta)$, second by $\sin(\beta)$ and the addition yields:

$$\frac{\partial k_s}{\partial \omega}\big(\cos(\alpha)\cos(\beta) - \sin(\alpha)\sin(\beta)\big) + \frac{\partial k_i}{\partial \omega}\big(\cos(\beta)^2 + \sin(\beta)^2\big) = 0$$

Using the Sine rule leads to:

$$\frac{\partial k_s}{\partial \omega}\cos(\Omega) + \frac{\partial k_i}{\partial \omega} = 0 \text{ with } \Omega = \alpha + \beta \quad \rightarrow \quad \upsilon_{gs} = \upsilon_{gi} \cdot \cos(\Omega)$$

with the group velocities $\upsilon_{gs} = \left(\dfrac{\partial k_s}{\partial \omega}\right)^{-1}$ and $\upsilon_{gi} = -\left(\dfrac{\partial k_i}{\partial \omega}\right)^{-1}$.

Matching of the group velocity of signal and idler in the direction of signal is therefore required to achieve broadband phase-matching.

A2) Birefringence of BBO, LBO and YCOB [138, 139]:

$$n[\theta_{pm}, \lambda_p] = \sqrt{\dfrac{1}{\dfrac{\cos^2[\theta_{pm}]}{n_o^2[\lambda_p]} + \dfrac{\sin^2[\theta_{pm}]}{n_e^2[\lambda_p]}}}, \quad n[\phi, \lambda_p] = n_Y[\lambda_p]\sqrt{\dfrac{1 + \tan^2[\phi]}{1 + \big(n_Y[\lambda_p]/n_X[\lambda_p]\big)^2 \tan^2[\phi]}}$$

BBO:
$$n_o^2 = 2.7359 + \frac{0.01878}{\lambda^2 - 0.01822} - 0.01354\lambda^2, \quad n_e^2 = 2.3753 + \frac{0.01224}{\lambda^2 - 0.01667} - 0.01516\lambda^2$$

LBO:
$$n_X^2 = 2.4542 + \frac{0.01125}{\lambda^2 - 0.01135} - 0.01388\lambda^2, \quad n_Y^2 = 2.5390 + \frac{0.01277}{\lambda^2 - 0.01189} - 0.01848\lambda^2$$

YCOB:
$$n_X^2 = 2.78232 + \frac{0.01120}{\lambda^2 - 0.08990} - 0.00007256\lambda^2,$$

$$n_Y^2 = 2.88739 + \frac{0.01223}{\lambda^2 - 0.08855} - 0.00006114\lambda^2$$

A3) Rate model for the P3HT population kinetics with conical intersection:

The transient optical signals shortly after the optical excitation are composed of the changing densities of the various species present in the sample and the spectrum of the extinction coefficient of each species. For clarity consider Fig. A3.

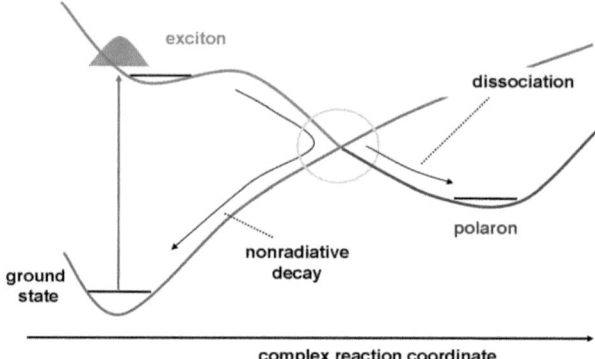

Fig. A3: Scheme of the exciton excitation, dissociation and nonradiative decay.

We denote the arial density of ground state, P3HT, exciton and postive polaron as n_{GS}, n_{exc} and n_{pol}. The changes after the optical excitation are given by

$$\frac{dn_{exc}}{dt} = -(k_{diss} + k_{nr})n_{exc} \tag{A3.1}$$

$$\frac{dn_{GS}}{dt} = +k_{nr}n_{exc} + k_{rec}n_{pol} \tag{A3.2}$$

$$\frac{dn_{pol}}{dt} = +k_{diss}n_{exc} - k_{rec}n_{pol}. \tag{A3.3}$$

k_{diss}, k_{nr} and k_{rec} are the first order rates for dissociation of excitons into polarons, the nonradiative decay of excitons down to the electronic ground state and the recombination of polarons to the ground state of P3HT. Since we are only concerned about the early signals and measurements at low excitation fluence, we can neglect the second term (charge recombination) in Eqns. A3.2 and A3.3 and any annihilation processes. The rate model can readily be solved to:

$$n_{exc}(t) = n_{exc}(t = 0^+) e^{-t(k_{diss} + k_{nr})} \tag{A3.4}$$

$$n_{GS}(t) = \left[n_{GS}^0 - n_{exc}(t=0^+)\right] +$$
$$+ \frac{k_{nr}}{k_{diss}+k_{nr}} n_{exc}(t=0^+)\left[1 - e^{-t(k_{diss}+k_{nr})}\right]$$
$$= \left[n_{GS}^0 - n_{exc}(t=0^+) + \frac{k_{nr}}{k_{diss}+k_{nr}} n_{exc}(t=0^+)\right] -$$
$$- \frac{k_{nr}}{k_{diss}+k_{nr}} n_{exc}(t=0^+) e^{-t(k_{diss}+k_{nr})} \quad (A3.5)$$

$$n_{pol}(t) = \frac{k_{diss}}{k_{diss}+k_{nr}} n_{exc}(t=0^+)\left[1 - e^{-t(k_{diss}+k_{nr})}\right]. \quad (A3.6)$$

We denote the ground state density before the excitation by n_{GS}^0 and the density of primary photoexcitations by $n_{exc}(t=0^+)$. Note that this quantity is directly related to n_{GS}

$$n_{GS}^0 - n_{exc}(t=0^+) = n_{GS}(t=0^+). \quad (A3.7)$$

Comparison of Eqns. A3.4 and A3.5 yields the important result that the ground state bleach has the same temporal evolution on the ultrafast time scale as the primary photoexcitation due to the nonvanishing nonradiative decay.

The transient optical signal is given by:

$$\Delta OD(t;\lambda) = \left[n_{GS}(t) - n_{GS}^0\right]\varepsilon_{GS}(\lambda) + n_{exc}(t)\varepsilon_{exc}(\lambda) + n_{pol}(t)\varepsilon_{pol}(\lambda). \quad (A3.8)$$

In the range below ~600 nm the extinction coefficient of the exciton and the polaron vanishes and the signal is dominated by the recovery of the GSB. In the NIR there is no ground state absorption and consequently also no GSB. The signal in the NIR is given by the population kinetics of the exciton and the polaron.

Appendix B1

Generation of sub-three-cycle, 16 TW light pulses by using noncollinear optical parametric chirped-pulse amplification

Daniel Herrmann, Laszlo Veisz, Raphael Tautz, Franz Tavella, Karl Schmid, Vladimir Pervak, and Ferenc Krausz

Reprinted with permission from
Optics Letters 34, 2459-24691 (2009).

DOI:10.1364/OL.34.002459

http://www.opticsinfobase.org/ol/abstract.cfm?URI=ol-34-16-2459

Copyright © 2009 The Optical Society of America

Generation of sub-three-cycle, 16 TW light pulses by using noncollinear optical parametric chirped-pulse amplification

Daniel Herrmann,[1,*] Laszlo Veisz,[1] Raphael Tautz,[1] Franz Tavella,[2] Karl Schmid,[1,3] Vladimir Pervak,[3] and Ferenc Krausz[1,3]

[1]*Max-Planck-Institut für Quantenoptik, Hans-Kopfermann-Strasse 1, 85748 Garching, Germany*
[2]*HASYLAB/DESY, Notkestrasse 85, 22607 Hamburg, Germany*
[3]*Department für Physik, Ludwig-Maximilians-Universität München, Am Coulombwall 1, 85748 Garching, Germany*
*Corresponding author: daniel.herrmann@mpq.mpg.de

Received May 1, 2009; revised July 16, 2009; accepted July 17, 2009;
posted July 21, 2009 (Doc. ID 110877); published August 11, 2009

We present a two-stage noncollinear optical parametric chirped-pulse amplification system that generates 7.9 fs pulses containing 130 mJ of energy at an 805 nm central wavelength and 10 Hz repetition rate. These 16 TW light pulses are compressed to within 5% of their Fourier limit and are carefully characterized by the use of home-built pulse diagnostics. The contrast ratio before the main pulse has been measured as 10^{-4}, 10^{-8}, and 10^{-11} at $t=-3.3$ ps, $t=-5$ ps, and $t=-30$ ps, respectively. This source allows for experiments in a regime of relativistic light–matter interactions and attosecond science. © 2009 Optical Society of America
OCIS codes: 190.4970, 190.7110, 140.4480, 190.4410, 320.7090, 140.3280.

Multiterawatt few-cycle light pulses offer the potential of pushing the frontiers of attosecond science [1,2], quantum coherent control [3], and laser-based particle acceleration [4]. Phase-controlled, intense, few-cycle light pulses can be used to generate powerful isolated attosecond pulses in the extreme UV (XUV) range [5,6], providing photon flux levels adequate for the autocorrelation of XUV pulses and XUV-pump–XUV-probe investigations [7]. For this purpose, carrier-envelope phase effects with few-cycle pulses have been demonstrated [8].

The invention of optical parametric chirped-pulse amplification (OPCPA) offered the prospect of generating few-cycle light pulses by providing sufficient gain-bandwidth to approach the terawatt level [9,10]. Noncollinear OPCPA (NOPCPA) offers many advantages over chirped pulse amplifiers (CPAs), such as a broad gain-bandwidth, high single-pass gain, wavelength tunability, low thermal effects, and a reduced stretching and compression factor [11]. Because of the many challenges, the amplification and compression of few-cycle, terawatt-class pulses were only recently demonstrated in the near-IR [12–15], and few-cycle pulses with less energy in the IR [16,17]. Few-cycle NOPCPA systems require accurate dispersion control during stretching and compression over a broad bandwidth, along with optimum phase-matching conditions, and a high-quality picosecond pump laser. This is needed to achieve an efficient conversion, to ensure a good spatial signal profile, to reach high pulse contrast, and to minimize optical parametric fluorescence (OPF), which can occur within the duration of the pump pulse. NOPCPA is also an efficient method to generate high-quality seed pulses in hybrid NOPCPA–CPA systems, which can eventually reach the petawatt level [18–21].

In this Letter, we report a NOPCPA system, which we call Light-Wave-Synthesizer-20 (LWS-20). It generates 7.9 fs, 130 mJ (16 TW) pulses at an 805 nm center wavelength and 10 Hz repetition rate with high contrast and an excellent beam profile for tight focusing.

The schematic of our LWS-20 NOPCPA setup is presented in Fig. 1. The setup starts with a Ti:sapphire oscillator (Rainbow, Femtolasers GmbH). Its central wavelength is 800 nm, and its bandwidth spans 350 nm with an output pulse energy of 4 nJ and a pulse duration of 5.5 fs at a 10 Hz repetition rate. 60% (2.4 nJ) of the oscillator output is used to generate the seed for the NOPCPA. The remaining 40% (1.6 nJ) is used to seed the flashlamp-pumped Nd:YAG pump laser amplifier (EKSPLA) providing all-optical synchronization.

The seed radiation for the pump amplifier, centered at 1064 nm with several picojoules of energy, is produced by soliton self-frequency shift in a photonic crystal fiber and is subsequently coupled into a regenerative Nd:YAG amplifier [22]. The output of the regenerative amplifier is split, each half being led to

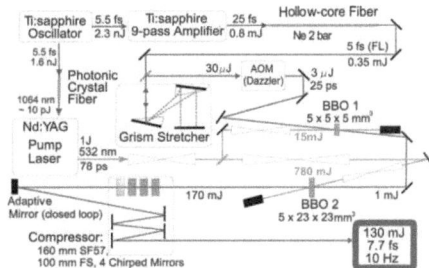

Fig. 1. Schematic layout of the NOPCPA setup.

a Nd:YAG amplifier chain consisting of a double- and two single-pass rods with increasing diameters (8, 12, and 18 mm, respectively). Via noncollinear Type II second-harmonic generation (SHG) in a DKDP crystal, both amplified beams are used to generate a single beam at 532 nm with a conversion efficiency of 45%, comprising pulses with 78 ps duration and 1 J of energy at a repetition rate of 10 Hz as pump for the NOPCPA. This beam is relay imaged onto two BBO crystals, which represent the NOPCPA stages.

In case of OPCPA, the contrast ratio between compressed pulse peak and OPF background strongly depends on the seed energy and on the OPCPA gain [23,24]. To enhance the pulse contrast, we use a new source to generate more energetic seed pulses compared with our previous work [13]. The oscillator output is first led through a 9-pass, 1 kHz Ti:sapphire CPA system (Femtopower Compact Pro, Femtolasers GmbH) and then coupled into a hollow-core fiber filled with neon gas at 2 bar (1500 Torr) absolute pressure, thereby achieving a broadened seed pulse spectrum ranging from 500 to 1000 nm. This spectrum corresponds to a 5 fs Fourier limit, and the pulse contains an energy of 0.35 mJ. We use a negative-dispersion reflection grism pair (10% efficiency) and an acousto-optic modulator (Dazzler, Fastlite, 10% transmission) to stretch the seed pulse to 25 ps with 3 µJ energy at a 10 Hz repetition rate for seeding the single-pass two-stage NOPCPA setup (phase-matching angle θ=23.62°, internal noncollinear angle α=2.23°). The first NOPCPA stage consists of a 5 mm×5 mm×5 mm Type I BBO crystal, is pumped by 15 mJ pulses at 532 nm with a peak intensity of 13 GW/cm^2, and amplifies the seed pulse to 1 mJ. The second NOPCPA stage is operated in saturation and consists of a 23 mm×23 mm×5 mm Type I BBO crystal and is pumped by 780 mJ pulses at 532 nm with a peak intensity of 8.2 GW/cm^2. After the second NOPCPA stage, the amplified signal pulse energy is 170 mJ with energy fluctuations within 3% rms (limited by 2% rms fluctuations of the pump energy) and negligible OPF. The pump-to-signal conversion is 22%.

Subsequently, the amplified signal beam is expanded from about 16 mm to 140 mm in diameter and compressed in bulk material consisting of 160-mm-long SF57 (Schott) and 100-mm-long fused silica. After the bulk compressor, the beam is down-collimated to a diameter of about 50 mm, sent to an adaptive mirror, and led into a vacuum compression chamber for final compression with four positive-dispersion chirped mirrors. The compressor has a calculated B integral below 1 and a throughput of 75% (including several silver mirrors). The pulse duration and residual phase are measured with a home-built all-reflective second-order single-shot autocorrelator and a home-built single-shot SHG frequency-resolved optical gating device (SHG-FROG). The autocorrelation trace is shown in Fig. 2(a) and reveals a 7.7 fs FWHM pulse duration with a deconvolution factor of 1.35, calculated from the spectrum. The shot-to-shot FWHM pulse duration fluctuations are measured

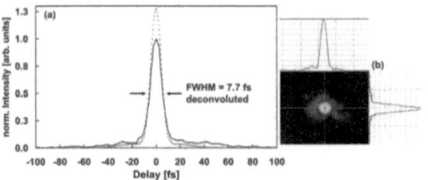

Fig. 2. (a) Second-order single-shot autocorrelation measurement (solid curve) and calculated autocorrelation trace, assuming a Fourier-limited pulse (dotted curve). (b) Focus of the amplified and compressed signal using an f=500 mm achromatic lens; the horizontal and vertical FWHM diameters are 10.58 and 10.25 µm, respectively.

with the single-shot autocorrelator to be within 4% rms. The SHG-FROG inversion results with removed time ambiguity are shown in Fig. 3. A pulse duration of 7.9 fs FWHM is measured while showing a fairly flat phase over the spectrum of the amplified pulse with a FROG error of 0.6%. The Fourier limit typically shows a duration of 7.5 fs FWHM. Therefore, compression is typically achieved to within 5% of the Fourier limit. The main pulse contains 83.5% (taken between the adjacent minima around the main peak)

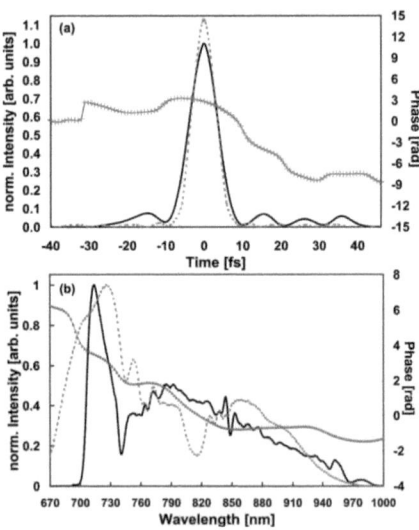

Fig. 3. FROG inversion results of the recompressed 130 mJ pulse (FROG error 0.6%): (a) temporal pulse intensity (solid curve) with removed time ambiguity, residual phase (crosses) and the Fourier-limited pulse (dotted curve). (b) Measured spectrum (696–994 nm, solid curve), retrieved residual phase (triangles), and unamplified stretched seed pulse spectrum (dotted curve).

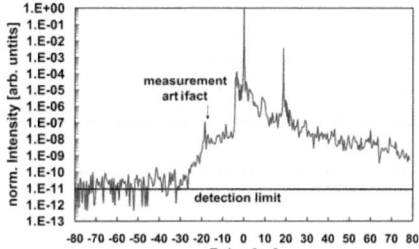

Fig. 4. (Color online) Third-order autocorrelation (negative delays correspond to the pulse leading edge); contrast of the amplified and compressed pulse; the peak at −18.8 ps is a measurement artifact belonging to the corresponding postpulse, and contributions for $t<-30$ ps are background noise.

of the total energy; in the case of a Fourier-limited pulse it is 94.7%. The contrast with respect to the main pulse is measured with an upgraded home-built third-harmonic generation autocorrelator (detection limit 10^{-11} [23]), and the results are shown in Fig. 4. The contrast ratio before the main pulse is measured as 10^{-4} at $t=-3.3$ ps, $\sim 10^{-8}$ at $t=-5$ ps, and $\sim 10^{-11}$ at $t=-30$ ps. We expect the contrast before $t=-30$ ps to be 10^{-12}–10^{-13}, calculated with a contrast of 10^{-7}–10^{-8} for the NOPCPA seed and a subsequent amplification factor of $\sim 10^5$. The postpulse at $t=18.8$ ps originates from a reflection inside the nine-pass amplifier, while the contribution between $t=-30$ ps and $t=-5$ ps was experimentally verified as a result of amplified spontaneous emission from the front end, which becomes amplified within the temporal window of the NOPCPA pump pulse.

Wavefront aberrations are corrected by using an adaptive mirror in a closed-loop configuration. The focus of the amplified and compressed super-Gaussian signal beam, using an $f=500$ mm achromatic lens, is shown in Fig. 2(b). The M^2 is measured as 2.7, and the Strehl ratio is 0.8.

In summary, we demonstrated for the first time, to our knowledge, the generation of 7.9 fs, 130 mJ (16 TW) light pulses at a central wavelength of 805 nm and a repetition rate of 10 Hz, achieved with our broadband LWS-20 NOPCPA system. This technology permits exploring attosecond physics and high-field interactions in a so far inaccessible parameter regime.

The authors are thankful to Jonas Kolenda (EKSPLA) for invaluable support with the pump laser and Xun Gu for useful discussions about FROG. This work was supported by Deutsche Forschungsgemeinschaft (contract TR18), by the association EURATOM-Max-Planck-Institut für Plasmaphysik, and by the Cluster of Excellence Munich Center for Advanced Photonics (MAP). D. Herrmann is also grateful to Studienstiftung des deutschen Volkes.

References
1. T. Brabec and F. Krausz, Rev. Mod. Phys. **72**, 545 (2000).
2. F. Krausz and M. Ivanov, Rev. Mod. Phys. **81**, 163 (2009).
3. H. Niikura, D. M. Villeneuve, and P. B. Corkum, Phys. Rev. A **73**, 021402(R) (2006).
4. K. Schmid, L. Veisz, F. Tavella, S. Benavides, R. Tautz, D. Herrmann, A. Buck, B. Hidding, A. Marcinkevičius, U. Schramm, M. Geissler, J. Meyer-ter-Vehn, D. Habs, and F. Krausz, Phys. Rev. Lett. **102**, 124801 (2009).
5. G. D. Tsakiris, K. Eidmann, J. Meyer-ter-Vehn, and F. Krausz, New J. Phys. **8**, 19 (2006).
6. E. Goulielmakis, M. Schultze, M. Hofstetter, V. S. Yakovlev, J. Gagnon, M. Uiberacker, A. L. Aquila, E. M. Gullikson, D. T. Attwood, R. Kienberger, F. Krausz, and U. Kleineberg, Science **320**, 1614 (2008).
7. Y. Nomura, R. Hörlein, P. Tzallas, B. Dromey, S. Rykovanov, Zs. Major, J. Osterhoff, S. Karsch, L. Veisz, M. Zepf, D. Charalambidis, F. Krausz, and G. D. Tsakiris, Nat. Phys. **5**, 124 (2009).
8. A. Baltuška, Th. Udem, M. Uiberacker, M. Hentschel, E. Goulielmakis, Ch. Gohle, R. Holzwarth, V. S. Yakovlev, A. Scrinzi, T. W. Hänsch, and F. Krausz, Nature **421**, 611 (2003).
9. A. Dubietis, G. Jonusauskas, and A. Piskarskas, Opt. Commun. **88**, 437 (1992).
10. I. N. Ross, P. Matousek, M. Towrie, A. J. Langley, and J. L. Collier, Opt. Commun. **144**, 125 (1997).
11. A. Dubietis, R. Butkus, and A. Piskarskas, IEEE J. Sel. Top. Quantum Electron. **12**, 163 (2006).
12. N. Ishii, L. Turi, V. S. Yakovlev, T. Fuji, F. Krausz, A. Baltuška, R. Butkus, G. Veitas, V. Smilgevičius, R. Danielius, and A. Piskarskas, Opt. Lett. **30**, 567 (2005).
13. F. Tavella, Y. Nomura, L. Veisz, V. Pervak, A. Marcinkevičius, and F. Krausz, Opt. Lett. **32**, 2227 (2007).
14. S. Witte, R. T. Zinkstok, A. L. Wolf, W. Hogervorst, W. Ubachs, and K. S. E. Eikema, Opt. Express **14**, 8168 (2006).
15. S. Adachi, N. Ishii, T. Kanai, A. Kosuge, J. Itatani, Y. Kobayashi, D. Yoshitomi, K. Torizuka, and S. Watanabe, Opt. Express **16**, 14341 (2008).
16. X. Gu, G. Marcus, Y. Deng, T. Metzger, C. Teisset, N. Ishii, T. Fuji, A. Baltuška, R. Butkus, V. Pervak, H. Ishizuki, T. Taira, T. Kobayashi, R. Kienberger, and F. Krausz, Opt. Express **17**, 62 (2009).
17. O. Chalus, P. K. Bates, M. Smolarski, and J. Biegert, Opt. Express **17**, 3587 (2009).
18. E. W. Gaul, M. Martinaz, J. Blakeney, A. Jochmann, M. Ringuette, D. Hammond, S. Marijanovic, R. Escamilla, and T. Ditmire, in Advanced Solid State Photonics, OSA Technical Digest Series (CD) (Optical Society of America, 2009), paper WD1.
19. C. Dorrer, I. A. Behishev, A. V. Okishev, J. D. Zuegel, Opt. Lett. **32**, 2143 (2007).
20. I. Jovanovic, C. A. Ebbers, and C. P. J. Barty, Opt. Lett. **27**, 1622 (2002).
21. H. Kiriyama, M. Mori, Y. Nakai, T. Shimomura, M. Tanoue, A. Akutsu, S. Kondo, S. Kanazawa, H. Okada, T. Motomura, H. Daido, T. Kimura, and T. Tajima, Opt. Lett. **33**, 645 (2008).
22. C. Y. Teisset, N. Ishii, T. Fuji, T. Metzger, S. Köhler, R. Holzwarth, A. Baltuška, A. M. Zheltikov, and F. Krausz, Opt. Express **13**, 6550 (2005).
23. F. Tavella, K. Schmid, N. Ishii, A. Marcinkevičius, L. Veisz, and F. Krausz, Appl. Phys. B **81**, 753 (2005).
24. F. Tavella, A. Marcinkevicius, and F. Krausz, New J. Phys. **8**, 219 (2006).

Appendix B2

Generation of 8 fs, 125 mJ Pulses Through Optical Parametric Chirped Pulse Amplification

Daniel Herrmann, Laszlo Veisz, Franz Tavella, Karl Schmid, Raphael Tautz, Alexander Buck, Vladimir Pervak and Ferenc Krausz

Reprinted with permission from

Advanced Solid-State Photonics, OSA Technical Digest Series (CD) (Optical Society of America, 2009), paper WA3.

http://www.opticsinfobase.org/abstract.cfm?URI=ASSP-2009-WA3

Copyright © 2009 The Optical Society of America

Generation of 8 fs, 125 mJ Pulses Through Optical Parametric Chirped Pulse Amplification

Daniel Herrmann[1], **Laszlo Veisz**[1], **Franz Tavella**[2], **Karl Schmid**[1,3], **Raphael Tautz**[1], **Alexander Buck**[1], **Vladimir Pervak**[3] **and Ferenc Krausz**[1,3]

[1] *Max-Planck Institut für Quantenoptik, Hans-Kopfermann-Strasse 1, 85748 Garching, Germany*
[2] *HASYLAB/DESY Notkestrae 85, 22607 Hamburg, Germany*
[3] *Department für Physik, Ludwig-Maximilian-Universität München, Am Coulombwall 1, 854748 Garching, Germany*
Author e-mail address: daniel.herrmann@mpq.mpg.de

Abstract: We report generation of three-cycle, 8 fs, 125 mJ optical pulses in a noncollinear optical parametric chirped-pulse amplifier (NOPCPA). These 16 TW laser pulses are compressed to within 6% of their Fourier limit.
© 2008 Optical Society of America
OCIS codes: 140.7090, 190.4970

High-peak-power few-cycle light sources are of major interest for a number of applications in nonlinear optics, highfield, and ultrafast science [1]. Few-cycle pulses open the route to generation, measurement and spectroscopic applications of isolated attosecond pulses and moreover steering the atomic-scale motion of electrons with controlled light fields [2–4]. Tsakiris *et al.* predict the emergence of an isolated single attosecond pulse when a solid density plasma surface is exposed to a multi-terawatt few-cycle driving laser pulse with cosine-shaped electrical field [5]. By using few-cycle laser systems, mono-energetic electron acceleration in the bubble regime is expected to show better reproducibility due to fewer parametric instabilities in the plasma [6]. Experimentally, Schmid *et al.* generated monoenergetic electrons in the bubble regime using 40 mJ, 8 fs pulses. These electrons showed up to 50 MeV energy and 5-10 mrad divergence (submitted).

The invention of OPCPA offered the prospect to generate few-cycle laser pulses by maintaining sufficient gain to approach the terawatt level for diverse center wavelengths [7]. But due to many challenges, only recently amplification and compression of few-cycle, terawatt-class pulses were demonstrated in the near-infrared [8–10], and few-cycle OPCPA systems with much less energy in the infrared [11]. Few-cycle OPCPA system requires accurate dispersion control during stretching and compression over broad bandwidth, optimum adjustment of the phase-matching parameters in the NOPCPA scheme, and a careful design of a short pulse pump laser to ensure good signal beam quality and to reduce optical parametric fluorescence (OPF, or superfluorescence) within the temporal window of the pump laser.

In this paper we report on an OPCPA system, that generates 8 fs, 125 mJ (16 TW) pulses at 805 nm center wavelength at 10 Hz repetition rate.

The schematic of our OPCPA setup is presented in Fig. 1. A key component in our system is the Ti:Sapphire oscillator (Femtolasers). The central wavelength is 800 nm and the bandwidth spans 350nm with an output pulse energy of 4 nJ and a pulse duration of 5.5 fs at 80 MHz repetition rate. 60% of the oscillator output is used to generate the seed of the OPCPA (2.4 nJ). 40% (1.6 nJ) is used to seed the flashlamp-pumped Nd:YAG pump laser amplifier (Ekspla) for optical synchronisation.

The seed radiation at 1064 nm with several pJ energy for the pump amplifier is produced by Raman-shift in a photonic crystal fiber and subsequently coupled into a regenerative Nd:YAG amplifier. The output of the regenerative amplifier is split, each half being led to a Nd:YAG amplifier chain consisting of double- and single-pass rods with increasing diameter (8 mm, 12 mm, 18 mm). Both amplifier chain outputs are temporally and spatially overlapped in a DKDP crystal for noncollinear type-II second-harmonic generation (SHG) leading to 80 ps pump pulses for the NOPCPA at 532 nm with 1.2 J energy.

The seed for the NOPCPA is first led through a 9-pass Ti:sapphire amplifier (Femtolasers) and then coupled into a neon-filled hollow-core fiber for providing a broadened seed pulse spectrum ranging from 475 nm up to 1020 nm with an energy of 0.4 mJ at 1 kHz repetition rate. After the hollow-core fiber crossed-polarized wave generation (XPW) is optionally used to improve the contrast by suppressing ASE and postpulses from the Ti:sapphire multipass amplifier before further amplification in the NOPCPA stages. This contrast improvement scheme shows 10% energy-efficiency and needs compressed pulses after the hollow-core fiber. We use a negative-dispersive reflection grism pair and an

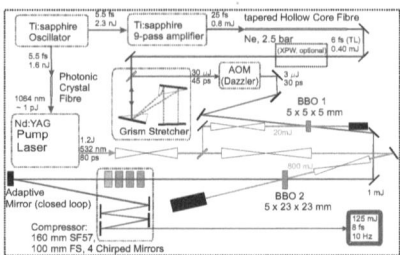

Fig. 1. Optical layout of the OPCPA experimental setup.

acousto-optic modulator (Dazzler, Fastlite) to stretch the seed pulse to 30 ps with $3\mu J$ energy at 10 Hz repetition rate for seeding the single-pass two-stage NOPCPA. For a detailed description of the stretcher see previous work [12].

In NOPCPA, the pump pulse is overlapped in time and space with the seed pulse while generating a so-called idler wave at the difference-frequency. By doing so, the seed pulse extracts energy from the pump pulse and becomes amplified. The parameters of the NOPCPA process, such as crystal length, phase-matching angle and noncollinear angle need to be carefully adjusted for efficient amplification of a broad part of the seed spectrum to support few-cycle, terawatt-class pulses. The first NOPCPA stage consists of a 5 mm x 5 mm x 5 mm type-I BBO crystal, is pumped by 15 mJ at 532 nm and amplifies the seed to 2 mJ. The second NOPCPA stage is operated in saturation and consists of a 23 mm x 23 mm x 5 mm type-I BBO crystal and pumped by 800-900 mJ at 532 nm. After the second NOPCPA stage, the amplified signal energy is 150 mJ with an energy fluctuation of 2.5% rms and negligible OPF.

Subsequently, the amplified signal pulse is expanded and compressed in bulks by using 160 mm SF57 and 100 mm fused silica. After the bulk compressor, the beam is downcollimated and led into a vacuum compression chamber for complete compression with 4 positive-dispersive chirped mirrors. The bulk compressor shows a B-integral below 1, a throughput of 84% and will soon be replaced to increase the transmission. The pulse duration and residual phase is measured with a home-built all-reflective second-order single-shot autocorrelator and a home-built single-shot second-harmonic generation frequency resolved optical gating device (SHG-FROG). The autocorrelation trace is shown in Fig. 2 and the SHG-FROG inversion results are shown in Fig. 3. Typically a pulse duration of 8 fs is achieved while showing a nearly flat phase over the spectrum of the amplified pulse. The Fourier-limit typically shows a duration of 7.6-8 fs FWHM. Compression is achieved to within 6% of the Fourier limit and the main pulse contains 80% of the total energy.

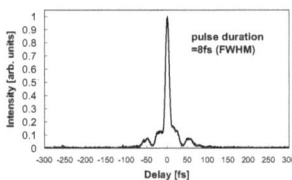

Fig. 2. Second-order single-shot autocorrelation measurement.

By using a deformable mirror in closed-loop configuration, wavefront aberrations are corrected. With further optimizing stretching and compression, we expect to increase the compressed spectral bandwidth (leading to a reduced Fourier-limit of 7.5 fs) while increasing the amplified energy after compression. We are working on the optimization of the contrast of the compressed pulse. Additionally, efforts to stabilize the carrier-envelope phase are under way.

In summary, we demonstrated for the first time to our knowledge the generation of 8 fs, 125 mJ (16 TW) light pulses at central wavelength of 805 nm and a repetition rate of 10 Hz from a broadband OPCPA system. This technology

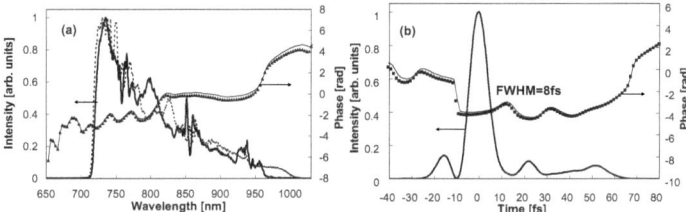

Fig. 3. FROG inversion results for the amplified pulse after compression (FROG error 0.6%): (a) measured pulse spectrum (solid curve), retrieved residual phase (triangles) and amplified spectrum prior to compression (dotted curve) with uncompressable spectral components above 980 nm, (b) temporal intensity of the recompressed 125 mJ pulse (solid curve) and residual phase (squares).

permits exploring attosecond physics and high-field interactions in a so far inaccessible parameter regime. An improved stretcher and compressor would already now lead to 20 TW few-cycle pulses. In principle, the OPCPA system is scalable even further to higher energies and lower pulse durations by using BBO crystals with larger aperture, higher pump energy, changing the NOPCPA parameters and implementing alternative NOPCPA schemes.

This work was supported by Deutsche Forschungsgemeinschaft (contract TR18), by the association EURATOM-Max-Planck Institut für Plasmaphysik and by the Cluster of Excellence Munich center for Advanved Photonics (MAP).

References

1. T. Brabec and F. Krausz, "Strong Field Laser Physics", Rev. Mod. Phys. **72**, 545 (2000).
2. A. Baltuska, Th. Udem, M. Uiberacker, M. Hentschel, E. Goulielmakis, Ch. Gohle, R. Holzwarth, V. S. Yakolev, A. Scrinzi, T. W. Hnsch, and F.Krausz, "Attosecond control of electronic processes by intense light fields ", Nature **421**,611 (2003).
3. R. Kienberger, E. Goulielmakis, M. Uiberacker, A. Baltuska, V. Yakovlev, F. Bammer, A. Scrinzi, Th. Westerwalbesloh, U. Kleineberg, U. Heinzmann, M. Drescher, F. Krausz, "Atomic transient recorder", Nature **427**, 817 (2004).
4. E. Goulielmakis, M. Schultze, M. Hofstetter, V. S. Yakovlev, J. Gagnon, M. Uiberacker, A. L. Aquila, E. M. Gullikson, D. T. Attwood, R. Kienberger, F. Krausz, U. Kleineberg, "Single-Cycle Nonlinear Optics", Science **320**, 1614 (2008).
5. G.D. Tsakiris, K. Eidmann, J. Meyer-ter-Vehn and F. Krausz, "Route to intense single attosecond pulses", New J. Phys. **8**, 19-39 (2006).
6. M. Geissler, J. Schreiber, and J. M. ter Vehn, "Bubble acceleration of electrons with few-cycle laser pulses", New J. Phys. **8**, 186 (2006).
7. A. Dubietis, G. Jonusauskas and A. Pisarskas, "Powerful femtosecond pulse generation by chirped and stretched pulse parametric amplification in BBO crystal", Opt. Commun. **88** 437-440 (1992).
8. F. Tavella, K. Schmid, N. Ishii, A. Marcinkevicius, L. Veisz, F. Krausz, "High-dynamic range pulse-contrast measurements of a broadband optical parametric chirped-pulse amplifier", Appl. Phys. B **3**, 753 (2005).
9. S. Witte, R.T. Zinkstok, A.L. Wolf, W. Hogervorst, W. Ubachs, K. S. E. Eikema, "A source of 2 terawatt, 2.7 cycle laser pulses based on noncollinear optical parametric chirped pulse amplification", Opt. Express **14**, 8168 (2006).
10. F. Tavella, A. Marcinkevicius, F. Krausz, "90 mJ parametric chirped pulse amplification of 10 fs pulses", Opt. Expr. **14**, 12822 (2006).
11. T. Fuji, N. Ishii, C. Y. Teisset, X. Gu, Th. Metzger, A. Baltuka, N. Forget, D. Kaplan, A. Galvanauskas, and F. Krausz "Parametric amplification of few-cycle carrierenvelope phase-stable pulses at 2.1μm", Opt. Lett. **31**, 1103 (2006).
12. F. Tavella, Y. Nomura, L. Veisz, V. Pervak, A. Marcinkevicius, F. Krausz, "Dispersion management for a sub-10-fs, 10 TW optical parametric chirped-pulse amplifier", Opt. Lett. **32**, 2227 (2007).

Appendix B3

Generation of Three-cycle, 16 TW Light Pulses by use of Optical Parametric Chirped Pulse Amplification

Daniel Herrmann, Laszlo Veisz, Raphael Tautz, Alexander Buck, Franz Tavella, Karl Schmid, Vladimir Pervak, Michael Scharrer, Philip Russell and Ferenc Krausz

Reprinted with permission from

CLEO/Europe and EQEC 2009 Conference Digest, (Optical Society of America, 2009), paper CG1_3.

http://www.opticsinfobase.org/abstract.cfm?URI=CLEO/Europe-2009-CG1_3

Copyright © 2009 The Optical Society of America

Generation of Three-cycle, 16 TW Light Pulses by use of Optical Parametric Chirped Pulse Amplification

Daniel Herrmann[1], Laszlo Veisz[1], Raphael Tautz[1], Alexander Buck[1], Franz Tavella[2], Karl Schmid[1,3], Vladimir Pervak[3], Michael Scharrer[4], Philip Russell[4] and Ferenc Krausz[1,3]

1. Max-Planck Institut für Quantenoptik, Hans-Kopfermann-Strasse 1, 85748 Garching, Germany
2. HASYLAB/DESY Notkestraße 85, 22607 Hamburg, Germany
3. Department für Physik, Ludwig-Maximilian-Universität München, Am Coulombwall 1, 854748 Garching, Germany
4. Max-Planck Institut für die Physik des Lichts, Günther-Scharowsky Str. 1/Bau 24, 91058 Erlangen, Germany

We report the first generation of three-cycle, 8 fs, 125 mJ optical pulses in a noncollinear optical parametric chirped-pulse amplifier (NOPCPA) operating at 805 nm central wavelength. These 16 TW laser pulses are compressed to within 5% of their Fourier limit. These parameters come in combination with an amplified pulse contrast-ratio of 10^{-5} at $\Delta t = \pm 5ps$ and $>10^{-7}$ at t<-5ps limited by the detection barrier. Numerical split-step simulations are performed to investigate potential improvements for the NOPCPA laser system.

Our laser system starts with a broad-bandwidth frontend with improved hollow-core fiber, which delivers seed pulses with an energy of 0.4 mJ at 1 kHz repetition rate. It is optically synchronized with a Nd:YAG laser, which provides 532 nm pump pulses of 1.2 J energy and 80 ps duration at 10 Hz repetition rate. We use a negative-dispersive grism pair and an acousto-optic modulator to stretch the seed pulse to 30 ps with 2 microjoules energy for seeding the single-pass two-stage NOPCPA [1]. After amplification, the signal energy is 150 mJ. Subsequently, the amplified signal pulse is compressed by use of glass bulks and chirped mirrors, and is characterized by using a home-built all-reflective single-shot second harmonic generation autocorrelator and frequency resolved optical gating device. The contrast is determined by use of a third-harmonic generation autocorrelator. 7 orders of magnitude contrast is measured in the ± 40 ps temporal window and many orders of magnitude better outside of this window [2]. See Fig. 1 for the experimental results. Recently, the crossed-polarized wave generation technique (XPW) [3] was implemented in the system leading to a significant improvement of the contrast, which is expected to be approximately 10 orders of magnitude 5 ps before the main pulse by suppressing the amplified spontaneous emission from the seed laser below the parametric superfluorescence level, which is simulated to be 10^{-10}. An improved detection limit is expected to resolve this high contrast in the near future. A Shack-Hartmann wavefront sensor and an adaptive mirror in a closed loop configuration are used to optimize the focusing properties of the laser to reach $>10^{19}$ W/cm^2 relativistic intensity on target. This laser system permits exploring attosecond and high-field physics in a so far inaccessible parameter regime.

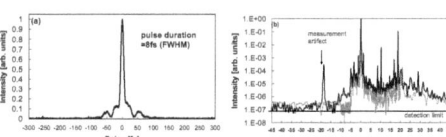

Fig. 1: A pulse duration of 8 fs ((a),deconvoluted) and a contrast-ratio of 10^{-5} at $\Delta t = \pm 5ps$ and $>10^{-7}$ at t<-5ps is measured ((b), contrast without (solid line) and with (dotted line) using XPW)

References

[1] F. Tavella, Y. Nomura, L. Veisz, V. Pervak, A. Marcinkevicius, F. Krausz, "Dispersion management for a sub-10-fs, 10 TW optical parametric chirped-pulse amplifier", Opt. Lett. **32**, 2227 (2007).

[2] F. Tavella, K. Schmid, N. Ishii, A. Marcinkevicius, L. Veisz, F. Krausz, "High-dynamic range pulse-contrast measurements of a broadband optical parametric chirped-pulse amplifier," Appl. Phys. B 81, 753 (2005).

[3] A. Jullien, O. Albert, F. Burgy, G. Hamoniaux, J. Rousseau, J. Chambaret, F. Augé-Rochereau, G. Chériaux, J. Etchepare, N. Minkovski, S. M. Saltiel, "10^{-10} temporal contrast for femtosecond ultraintense lasers by cross-polarized wave generation," Opt. Lett. 30, 920 (2005).

Appendix B4

Investigation of two-beam-pumped noncollinear optical parametric chirped-pulse amplification for the generation of few-cycle light pulses

Daniel Herrmann, Raphael Tautz, Franz Tavella, Ferenc Krausz, and Laszlo Veisz

Reprinted with permission from
Optics Express 18, 4170-4183 (2010).

DOI: 10.1364/OE.18.004170

http://www.opticsinfobase.org/oe/abstract.cfm?URI=oe-18-5-4170

Copyright © 2010 The Optical Society of America

Investigation of two-beam-pumped noncollinear optical parametric chirped-pulse amplification for the generation of few-cycle light pulses

Daniel Herrmann,[1,2,3,4*] Raphael Tautz,[1,2,3,5] Franz Tavella,[6]
Ferenc Krausz,[2,3] and Laszlo Veisz[2]

[1] *These two authors contributed equally to this work.*
[2] *Max-Planck-Institut für Quantenoptik, Hans-Kopfermann-Strasse 1, 85748 Garching*
[3] *Department für Physik, Ludwig-Maximilians-Universität München, Am Coulombwall 1, 85748 Garching, Germany*
[4] *Present address: Lehrstuhl für BioMolekulare Optik, Department für Physik, Ludwig-Maximilians-Universität, Oettingenstrasse 67, 80538 München, Germany*
[5] *Present address: Photonics and Optoelectronics Group, Department of Physics and Center for Nanoscience, Ludwig-Maximilians-Universität, Amalienstr. 54, 80799 München, Germany*
[6] *HASYLAB/DESY, Notkestrasse 85, 22607 Hamburg, Germany*
*Corresponding author: d.herrmann@physik.uni-muenchen.de

Abstract: We demonstrate a new and compact ϕ-plane-pumped noncollinear optical parametric chirped-pulse amplification (NOPCPA) scheme for broadband pulse amplification, which is based on two-beam-pumping (TBP) at 532 nm. We employ type-I phase-matching in a 5 mm long BBO crystal with moderate pump intensities to preserve the temporal pulse contrast. Amplification and compression of the signal pulse from 675 nm - 970 nm is demonstrated, which results in the generation of 7.1-fs light pulses containing 0.35 mJ energy. In this context, we investigate the pump-to-signal energy conversion efficiency for TBP-NOPCPA and outline details for few-cycle pulse characterization. Furthermore, it is verified, that the interference at the intersection of the two pump beams does not degrade the signal beam spatial profile. It is theoretically shown that the accumulated OPA phase partially compensates for wave-vector mismatch and leads to extended broadband amplification. The experimental outcome is supported by numerical split-step simulations of the parametric signal gain, including pump depletion and parametric fluorescence.

© 2010 Optical Society of America

OCIS codes: (190.7110) Ultrafast nonlinear optics; (190.4970) Parametric oscillators and amplifiers; (190.4975) Parametric processes; (320.7100) Ultrafast measurements; (230.4320) Nonlinear optical devices; (260.7120) Ultrafast phenomena

References and links

1. D. Polli, M. R. Antognazza, D. Brida, G. Lanzani, G. Cerullo, and S. De Silvestri, "Broadband pump-probe spectroscopy with sub-10-fs resolution for probing ultrafast internal conversion and coherent phonons in carotenoids," Chem. Phys. **350**, 45-55 (2008).
2. T. Brabec and F. Krausz, "Intense few-cycle laser fields: Frontiers of nonlinear optics," Rev. Mod. Phys. **72**, 545-591 (2000).
3. F. Krausz and M. Ivanov, "Attosecond physics," Rev. Mod. Phys. **81**, 163-234 (2009).

4. K. Schmid, L. Veisz, F. Tavella, S. Benavides, R. Tautz, D. Herrmann, A. Buck, B. Hidding, A. Marcinkevičius, U. Schramm, M. Geissler, J. Meyer-ter-Vehn, D. Habs, and F. Krausz, "Few-Cycle Laser-Driven Electron Acceleration," Phys. Rev. Lett. **102**, 124801 (2009).
5. G. D. Tsakiris, K. Eidmann, J. Meyer-ter-Vehn, and F. Krausz, "Route to intense single attosecond pulses," New J. Phys. **8** (2006) 19.
6. A. Baltuška, Th. Udem, M. Uiberacker, M. Hentschel, E. Goulielmakis, Ch. Gohle, R. Holzwarth, V. S. Yakolev, A. Scrinzi, T. W. Hänsch, and F. Krausz, "Attosecond control of electronic processes by intense light fields," Nature **421**, 611-615 (2003).
7. R. Hörlein, Y. Nomura, D. Herrmann, M. Stafe, I. B. Földes, S. G. Rykovanov, F. Tavella, A. Marcinkevičius, F. Krausz, L. Veisz, and G. D. Tsakiris, "Few-cycle harmonic emission from solid density plasmas," (in preparation).
8. A. Dubietis, G. Jonusauskas, and A. Piskarskas, "Powerful femtosecond pulse generation by chirped and stretched pulse parametric amplification in BBO crystal," Opt. Commun. **88**, 437-440 (1992).
9. I. N. Ross, P. Matousek, G. H. C. New, and K. Osvay, "Analysis and optimization of optical parametric chirped pulse amplification," J. Opt. Soc. Am. B **19**, 2945-2956 (2002).
10. A. Dubietis, R. Butkus, and A. Piskarskas, "Trends in chirped pulse optical parametric amplification," IEEE J. Sel. Top. Quantum Electron. **12**, 163-172 (2006).
11. S. Witte, R. T. Zinkstok, A. L. Wolf, W. Hogervorst, W. Ubachs, and K. S. E. Eikema, "A source of 2 terawatt, 2.7 cycle laser pulses based on noncollinear optical parametric chirped pulse amplification," Opt. Express **14**, 8168-8177 (2006).
12. F. Tavella, Y. Nomura, L. Veisz, V. Pervak, A. Marcinkevičius, and F. Krausz, "Dispersion management for a sub-10-fs, 10 TW optical parametric chirped-pulse amplifier," Opt. Lett. **32**, 2227-2229 (2007).
13. D. Herrmann, L. Veisz, R. Tautz, F. Tavella, K. Schmid, V. Pervak, and F. Krausz, "Generation of sub-three-cycle, 16 TW light pulses by using noncollinear optical parametric chirped-pulse amplification," Opt. Lett. **34**, 2459-2461 (2009).
14. A. Dubietis, R. Danielius, G. Tamošauskas, and A. Pisarskas, "Combining effect in a multiple-beam-pumped optical parametric amplifier," J. Opt. Soc. Am. B **15**, 1135-1139 (1998).
15. A. Marcinkevičius, A. Piskarskas, V. Smilgevičius, and A. Stabinis, "Parametric superfluorescence excited in a nonlinear crystal by two uncorrelated pump beams," Opt. Commun. **158**, 101-104 (1998).
16. D. Brida, G. Cirmi, C. Manzoni, S. Bonora, P. Villoresi, S. De Silvestri, and G. Cerulo, "Sub-two-cycle light pulses at $1.6\mu m$ from an optical parametric amplifier", Opt. Lett. **33**, 741-743 (2008).
17. T. S. Sosnowski, P. B. Stephens, and T. B. Norris, "Production of 30-fs pulses tunable throughout the visible region by a new technique in optical parametric amplification," Opt. Lett. **21**, 140-142 (1996).
18. E. Žeromskis, A. Dubietis, G. Tamošauskas, A. Piskarskas, "Gain bandwidth broadening of continuum-seeded optical parametric amplifier by use of two pump beams," Opt. Comm. **203**, 435-440 (2002).
19. G. Tamošauskas, A. Dubietis, G. Valiulis, and A. Piskarskas, "Optical parametric amplifier pumped by two mutually incoherent laser beams," Appl. Phys. B **91**, 305307 (2008).
20. C. Wang, Y. Leng, B. Zhao, Z. Zhang, Z. Xu, "Extremely broad gain spectra of two-beam-pumped optical parametric chirped-pulse amplifier," Opt. Commun. **237**, 169-177 (2004).
21. R. L. Sutherland, 1996, *Handbook of Nonlinear Optics* (Marcel Dekker, New York).
22. G. Cerullo and S. de Silvestri, "Ultrafast optical parametric amplifiers," Rev. of Sci. Instr. **74**, 1-18 (2003).
23. V. G. Dmitriev, G. G. Gurzadyan, and D. N. Nikogosyan, *Handbook of Nonlinear Optical Crystals* (Springer, New York).
24. J. A. Armstrong, N. Bloembergen, J. Ducuing, and P. S. Pershan, "Interactions between light waves in a nonlinear dielectric," Phys. Rev. **127**, 1918-1939 (1962).
25. R. A. Baumgartner and R. L. Byer, "Optical Parametric Amplification," IEEE J. of. Quant. Electr. **QE-15**, 432-444 (1979).
26. F. Tavella, A. Marcinkevičius, and F. Krausz, "Investigation of the superfluorescence and signal amplification in an ultrabroadband multiterawatt optical parametric chirped pulse amplifier system," New J. Phys. **8** (2006) 219.
27. A. Dement'ev, O. Vrublevskaja, V. Girdauskas, and R. Kazragyte, "Numerical Analysis of Short Pulse Optical Parametric Amplification Using Type I Phase Matching," Nonl. Anal.: Modelling and Control **9**, 39-53 (2004).
28. A. Kurtinaitis, A. Dementjev, and F. Ivanauskas, "Modeling of Pulse Propagation factor Changes in Type II Second-Harmonic Generation," Nonl. Anal.: Modelling an Control **6**, 51-69 (2001).
29. G. A. Reider, 1997, *Photonik* (Springer, New York).
30. L. Hongjun, Z. Wei, C. Guofu, W. Yishan, C. Zhao, and R. Chi, "Investigation of spectral bandwidth of optical parametric amplification," Appl. Phys. B **79**, 569576 (2004).
31. F. Tavella, K. Schmid, N. Ishii, A. Marcinkevičius, L. Veisz, and F. Krausz, "High-dynamic range pulse-contrast measurements of a broadband optical parametric chirped-pulse amplifier," Appl. Phys. B **3**, 753 (2005).
32. G. Arisholm, "General numerical methods for simulating second-order nonlinear interactions in birefringent media," J. Opt. Soc. Am. B **14**, 2543-2549 (1997).
33. S. Witte, R. T. Zinkstok, W. Hogervorst, and K. S. E. Eikema, "Numerical simulation for performance optimization of a few-cycle terawatt NOPCPA system," Appl. Phys. B **87**, 677684 (2007).

34. A. Picozzi and M. Haeltermann, "Influence of walk-off, dispersion, and diffraction on the coherence of parametric fluorescence," Phys. Rev. E, **63**, 056611-1 - 056611-11 (2001).
35. A. Gatti, H. Wiedemann, L. A. Lugitao, and I- Marzoli, "Langevin treatment of quantum fluctuations and optical patterns in optical parametric oscillators below threshold," Phys. Rev. A, **56**, 877-897 (1997).
36. D. A. Kleinmann, "Theory of Optical Parametric Noise," Phys. Rev., **174**, 1027-1041 (1968).
37. F. Salin, P. Georges, G. Roger, and A. Brun, "Single-shot measurement of a 52-fs pulse," Applied Optics **26**, 4528-4531 (1987).
38. A. Brun, P. Georges, G. Le Saux, and F. Salin, "Single-shot characterization of ultrashort light pulses," J. Phys. D: Appl. Phys. **24**, 1225-1233 (1991).
39. H. Mashiko, A. Suda, and K. Midorikawa, "All-reflective interferometric autocorrelator for the measurement of ultra-short optical pulses," Appl. Phys. B **76**, 525-530 (2003).
40. R. Trebino, 2000, *Frequency-Resolved Optical Gating: The Measurement of Ultrashort Laser Pulses* (Kluwer Academic Publishers, Norwell, USA).

1. Introduction

Light pulses with duration of only a few optical cycles open up novel parameter regimes for a number of applications, ranging from time-resolved optical spectroscopy to high-field science [1, 2]. Progress in strong-field physics such as attosecond science [3] or laser-based particle acceleration [4] depend on the availability of suitable laser systems. Phase-controlled, intense, high-contrast, few-cycle light pulses can be used to generate powerful isolated attosecond pulses in the XUV range [5]. For this purpose, carrier-envelope phase effects with few-cycle pulses have been demonstrated [6, 7].

The invention of OPCPA offered the prospect of generating few-cycle laser pulses by providing sufficient gain and gain-bandwidth to approach the terawatt level [8, 9]. NOPCPA offers many advantages over chirped pulse amplifiers (CPA), such as a broad gain-bandwidth, high single-pass gain, wavelength tunability and low thermal effects [10]. Due to many challenges, amplification and compression of few-cycle, terawatt-class pulses have only recently been demonstrated in the near-infrared [11, 12, 13]. Few-cycle NOPCPA systems require accurate dispersion control during stretching and compression over a broad bandwidth, along with optimum phase-matching conditions, and a high quality picosecond pump laser. This is needed to achieve an efficient conversion, to ensure a good spatial signal profile and to provide a high temporal pulse contrast. It was shown, that one can minimize optical parametric fluorescence (OPF) and reach a high pulse contrast suitable for high-field physics via moderately pumped NOPCPA in combination with a strong seed in 5 mm long BBO type-I phase-matching [13]. OPCPA and OPA allow for the use of multiple pump beams, which is an important but so far unexploited feature for laser systems [10]. The reason for this promising attribute is, that the relative pump phases are not important and no interferometric phase alignment is needed since the phase difference between the signal and the pump pulses are taken away by the idler pulse, hence compensating for random differences between them [14, 15]. Neighboring spectral regions of the seed can be amplified by using multiple pump beams in individual phase-matching geometry. Thus, a broader amplified signal spectrum can be achieved compared to employing just one pump beam. TBP can also increase the repetition rate for high-power OPCPA systems. Alternative approaches to TBP-NOPCPA for the generation of 1-3 cycle pulses are one-beam-pumping (OBP) collinear OPCPA at degeneracy and cascaded NOPCPA employing angular detuning [16, 17]. Previous TBP experimental studies were exclusively done on narrowband NOPCPA without revealing all characteristics like conversion efficiency, broadband signal spectra, broadband signal compressibility, pump beam interference, signal beam quality and simulated gain (based on five nonlinear coupled wave equations including pump depletion and OPF) at the same time [14, 18, 19]. Theoretical broadband TBP-NOPCPA studies predicted broadband compression to be very challenging because of phase jumps due to the TBP scheme [20]. Moreover, TBP-NOPCPA needs to be investigated in detail before implementation of this scheme in multi-terawatt laser systems [13]. This is the motivation for the present publication.

2. TBP-NOPCPA scheme and numerical simulation

NOPCPA involves phase-matching of the pump, signal and idler waves inside of a birefringent nonlinear optical crystal showing a second-order suceptibility. Phase-matching can be achieved by adjusting the phase-matching angles θ_{pm}, Φ and the noncollinear angle σ aiming for the minimum wave-vector mismatch [9, 21, 22]:

$$\Delta \mathbf{k}_l = \mathbf{k}_{pl} - \mathbf{k}_s - \mathbf{k}_{il} \quad \forall\, l = 1, 2, \tag{1}$$

where the index l separates the different pump and idler pulses. BBO is the favored nonlinear crystal material because of its high damage threshold, large nonlinearity and advantageous optical properties for broadband phase-matching [23].

OPCPA is described by the nonlinear coupled wave equations [21, 24, 25, 26] which need to be extended for describing TBP-NOPCPA:

$$\frac{\delta A_{s1}}{\delta z} + \sum_m \alpha_{s1m} \frac{\delta^m A_{s1}}{\delta t^m} + \frac{j\beta_{s11}}{r}\frac{\delta}{\delta r}\left(r\frac{\delta A_{s1}}{\delta r}\right) + \ldots + \gamma_{s1}\frac{\delta A_{s1}}{\delta r}$$
$$= -j\kappa_{s1} A_{i1}^* A_{p1} e^{-j\Delta k_1 z} - j\kappa_{s2} A_{i2}^* A_{p2} e^{-j\Delta k_2 z} + \sqrt{\varepsilon_{s1}}\xi_{s1}(z,t) \tag{2}$$

$$\frac{\delta A_{i1}}{\delta z} + \sum_m \alpha_{i1m}\frac{\delta^m A_{i1}}{\delta t^m} + \frac{j\beta_{i11}}{r}\frac{\delta}{\delta r}\left(r\frac{\delta A_{i1}}{\delta r}\right) + \ldots + \gamma_{i1}\frac{\delta A_{i1}}{\delta r} = -j\kappa_{i1}A_s^* A_{p1} e^{-j\Delta k_1 z} + \sqrt{\varepsilon_{i1}}\xi_{i1}(z,t) \tag{3}$$

$$\frac{\delta A_{i2}}{\delta z} + \sum_m \alpha_{i2m}\frac{\delta^m A_{i2}}{\delta t^m} + \frac{j\beta_{i21}}{r}\frac{\delta}{\delta r}\left(r\frac{\delta A_{i2}}{\delta r}\right) + \ldots + \gamma_{i2}\frac{\delta A_{i2}}{\delta r} = -j\kappa_{i2}A_s^* A_{p2} e^{-j\Delta k_2 z} + \sqrt{\varepsilon_{i2}}\xi_{i2}(z,t) \tag{4}$$

$$\frac{\delta A_{p1}}{\delta z} + \sum_m \alpha_{p1m}\frac{\delta^m A_{p1}}{\delta t^m} + \frac{j\beta_{p11}}{r}\frac{\delta}{\delta r}\left(r\frac{\delta A_{p1}}{\delta r}\right) + \ldots + \gamma_{p1}\frac{\delta A_{p1}}{\delta r} = -j\kappa_{p1}A_s A_{i1} e^{j\Delta k_1 z} \tag{5}$$

$$\frac{\delta A_{p2}}{\delta z} + \sum_m \alpha_{p2m}\frac{\delta^m A_{p2}}{\delta t^m} + \frac{j\beta_{p21}}{r}\frac{\delta}{\delta r}\left(r\frac{\delta A_{p2}}{\delta r}\right) + \ldots + \gamma_{p2}\frac{\delta A_{p2}}{\delta r} = -j\kappa_{p2}A_s A_{i2} e^{j\Delta k_2 z} \tag{6}$$

where $\alpha_{s1m}, \alpha_{ilm}$ and α_{plm} are the dispersion coefficients with order m. β_{s11}, β_{il1} and β_{pl1} are the diffraction coefficients at first order and (...) are the higher order terms [27, 28]. γ_{s1}, γ_{il} and γ_{pl} are the spatial overlap coefficients due to the noncollinear geometry. κ_{sl}, κ_{il} and κ_{pl} are the nonlinear coupling coefficients as a function of signal wavelength [29], Δk_l is the wave-vector mismatch given by Eq. (1) and z is the direction of propagation.

If there is no input idler present ($A_{i1} = A_{i2} = 0$), the initial idler phase Φ_{il} self-adjusts to achieve maximum initial pump to signal and idler conversion efficiency for each pump-signal interaction separately. This leads to the generalized phase $\Omega(z = 0)$ at the crystal input:

$$\Omega(0) = \Phi_{pl}(0) - \Phi_s(0) - \Phi_{il}(0) = -\frac{\pi}{2} \quad \forall\, l = 1, 2. \tag{7}$$

In the prescence of wavevector mismatch $\Delta \mathbf{k}_l$, there is an accumulated phase $|\Delta \mathbf{k}_l L|$ with propagation distance. Consequently, the phases of the waves have to change to maintain the $\Omega = -\pi/2$ criteria (Eq. (7)). This means, that an additional phase will be imprinted on the signal wave during amplification. The phase terms $\Phi_{p1,p2,s,i1,i2}$ for the pump waves, signal wave and idler waves can be obtained by solving the imaginary part of the coupled wave equations (Eq. (2-6)) and using the Manley-Rowe-relation. If the input signal intensity is small compared to the input pump intensity, i.e. in case of low pump depletion, one can obtain a good approximation for the signal phase at the output for each isolated pump-signal interaction separately (i.e. one pump beam is blocked at this time) [9]:

$$\Phi_{sl}(z=L) = \Phi_s(0) - \frac{\Delta k_l L}{2} + arctan\{\frac{\Delta k_l \times tanh[[g_l^2 - (\Delta k_l/2)^2]^{1/2}L]}{2[g_l^2 - (\Delta k_l/2)^2]^{1/2}}\} \quad \forall l = 1,2, \quad (8)$$

where $g_l = 4\pi d_{eff}\sqrt{I_{pl}(0)/(2\varepsilon_0 n_{pl} n_s n_{il} c \lambda_s \lambda_{il})}$ is the gain coefficient and

$$d_{eff} = d_{31} sin\theta_{pm} - d_{22} cos\theta_{pm} sin3\phi_l \quad \forall l = 1,2 \quad (9)$$

is the effective nonlinear coefficient. Since the accumulated signal phase Φ_{sl} depends on the pump intensity, a shot-to-shot variation of the carrier-envelope phase (CEP) is possible in the parametric amplifier, if the pump intensity is not constant in time. A maximum phase-slippage of π over the crystal length L is acceptable for a coherent built-up of the amplified signal in the small signal gain regime [22, 30]. This defines a criteria for the parametric bandwidth and leads to:

$$|\Delta \mathbf{k}_l L| \lesssim \pi \quad \forall l = 1,2. \quad (10)$$

We choose the TBP-NOPCPA geometry shown in Fig.2(a), to ensure optimum type-I phase-matching. In this setup, the critical phase-matching angle θ_{pm} can be kept the same for both pump-signal interactions, whereas the internal noncollinear angle σ_l is adjusted for both interactions individually to achieve the broadest amplified signal spectrum. The two pump beams are located in the ϕ-plane, which is perpendicular to the critical phase-matching plane (θ-plane). The change of d_{eff} due to the deviation of the non-critical phase-matching angle ϕ_l for the individual interactions is negligible but nevertheless was taken into account for the numerical simulations.

This TBP-NOPCPA setup is essentially different from conventional type-I NOPCPA geometries, where the so-called tangential phase-matching or the Poynting vector walk-off compensation phase-matching are used. In these cases, the pump and signal beam are located in the θ-plane. In this conventional configuration, it would not be possible to achieve a TBP geometry, where the two separate pump beams have the same phase-matching angle θ_{pm} and the noncollinear angles σ_l can be changed without changing θ_{pm}.

We perform symmetrized numerical split-step simulations of NOPCPA [26, 28, 32, 33] with the extension to the TBP scheme and including quantum noise field terms $\varepsilon_{s1,il}\xi_{s1,il}(z,t)$ into the coupled wave equations (Eq. (2-6)) [34]. The complex stochastic variables $\xi_{s1,il}(z,t)$ have a Gaussian distribution with a zero mean value $\langle \xi_{s,il}(z,t) \rangle = 0$ and the correlation $\langle \xi_q(z,t)\xi_r(z',t') \rangle = \delta_{q,r}\delta(z-z')\delta(t-t')$ [35]. $\varepsilon_{s1,il}$ are the noise intensities of the signal and the idler waves. The spatial overlap coefficient for the signal γ_{s1} is set to zero as the signal propagates normal to the crystal plane. We perform the parametric amplification in the time domain within the so-called *nonlinear step* (Eq. (2-6)) for the amplitudes of the signal wave, the OPF in the signal direction, the idler waves, the OPF in the idler directions and the pump waves. For the pump pulses we employ the measured beam shapes: a 6th-order super-Gasussian beam in space and a Gaussian pulse in time. In case of the seed, we use a Gaussian pulse in time and space as approximation. As further input parameters, we take measured seed and pump characteristics prior to amplification. We choose an unamplified seed spectrum ranging from 650 nm until 1060 nm in the simulation to reveal the maximum parametric gain bandwidth. The temporal delays between seed pulse and each pump pulse are chosen to achieve best match between simulation results and experimental outcome. For the broadest signal spectrum, we choose $\sigma_1 = 2.22°, \sigma_2 = -2.16°$ and $\theta_{pm} = 23.62°$. The wavevector mismatch Δk is assumed to be zero only for the OPF. This approximation is considered valid, since the transition rate from the pump photon to signal and idler photon is highest in the direction of smallest phase-mismatch and consequently only in this case efficient amplification of OPF is possible from the quantum noise level [26, 36].

Dispersion, diffraction and spatial overlap due to the noncollinearity for the TBP-NOPCPA process is taken into account in the *linear step*. For the dispersion we expand the spectral phase $\Delta \varphi_{s1,pl,il}(\omega)$ in a Taylor series until 4th order, including group velocity mismatch, and apply it in the frequency domain involving the Fourier transform $\mathscr{F}$:

$$A_{s1,pl,il}(r,z+\Delta z,t) = \mathscr{F}^{-1}\{\mathscr{F}[A_{s1,pl,il}(z,t)]e^{j\Delta \varphi_{s1,pl,il}(\omega)}\} \quad \forall l = 1,2. \tag{11}$$

For the diffraction (diff) and spatial overlap due to the noncollinearity (nc), we include a spatial phase term $\Delta \varphi_{s1,pl,il}^{spatial} = \Delta \varphi_{s1,pl,il}^{nc} + \Delta \varphi_{s1,pl,il}^{diff}$ by applying the linear Fourier step to the transposed data matrices [26, 27]:

$$A_{s1,pl,il}^{\dagger}(r,z+\Delta z,t) = \mathscr{F}^{-1}\{\mathscr{F}[A_{s1,pl,il}^{\dagger}(z,t)]e^{j\Delta \varphi_{s1,pl,il}^{spatial}}\} \quad \forall l = 1,2. \tag{12}$$

We perform the numerical model, based on Eq. (2-6) and Eq. (11), (12) with 500 split-step iteration loops for a 5 mm long type-I BBO crystal.

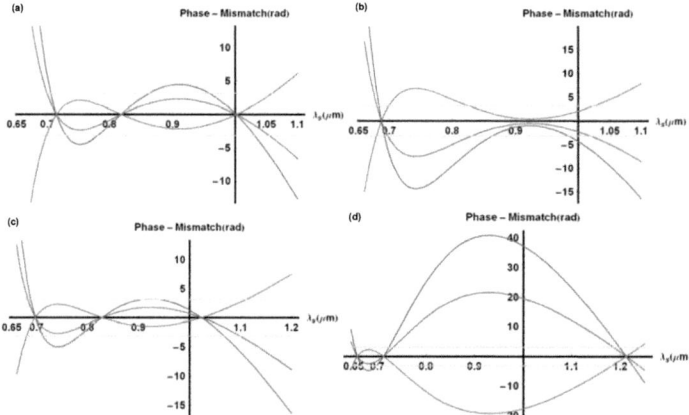

Fig. 1. Wave-vector mismatch $\Delta k_l L$ (blue), OPA-phase $\Phi_{sl}(z=L)$ (red), sum of both (purple) and $\Delta k_l L = \pm \pi$ (green) for the cases of (a) $\sigma = 2.22°$, $\theta_{pm} = 23.62°$, L=5 mm, λ_p=532 nm (b) $\sigma = 2.16°$, $\theta_{pm} = 23.62°$, L=5 mm, λ_p=532 nm (c) $\sigma = 2.155°$, $\theta_{pm} = 23.55°$, L=3 mm, λ_p=532 nm and (d) $\sigma = 2.04°$, $\theta_{pm} = 23.62°$, L=5 mm, λ_p=515 nm

To directly show, outside the numerical simulations, which spectral components can be amplified in TBP-NOPCPA employing 5 mm long BBO type-I phase-matching, we calculate the wave-vector mismatch by using Eq. (1) and the signal phase through Eq. (8) separately for each pump-signal interaction (i.e. while the other pump beam is blocked). The results are shown in Fig. 1(a) and (b). According to Eq. (10), the spectral components between 675 nm and 1050 nm can be amplified in case of TBP-NOPCPA (Fig. 1(a) and (b)) and between 700 nm and 1050 nm in case of broadband OBP-NOPCPA (Fig. 1(a)). This is in agreement with the measured signal spectra shown in Fig. 4, where the unamplified seed spectrum ranges from 650 nm until 970 nm. Moreover, Fig. 1 shows, that the signal phase Φ_{sl} accumulated over the crystal length L partially compensates for wave-vector mismatch $\Delta k_l L$ and pushes the overall phase-mismatch for the amplified spectral components below the limit in Eq. (10) for each individual pump-signal interaction. Maximum amplification is achieved, if the criterion for the general phase (Eq. (7)) is fulfilled. If there is no idler present at the beginning of the amplification, the idler phase self-adjusts to maximize the gain. During amplification however, the wave-vector mismatch (Eq. (1)) forces the phases of the waves to maintain this criterion. Nevertheless,we want to emphasize, that it remains to be theoretically investigated, how the total wave-vector mismatch and the total signal phase evolve, if both pump beams are present in the crystal. Nevertheless, the amplified spectra in Fig. 4 show, that the latter unknown effect does not noticeable change the phase-matched bandwidth compared to taking the phase-mismatch for each pump-signal interactions separately and applying the criterion for amplification in Eq. (10) afterwards. This approach is even exact for spectral regions in Fig. 4(b), where amplification is only due to one pump beam; i.e. for $\leq\sim$ 700 nm and for $\sim$700 nm-800 nm.

In principle, such a broad spectrum from 675 nm to $\geq$1050 nm can also be amplified with a similar gain using a 3 mm long BBO crystal in OBP-NOPCPA geometry, pumped with about 3 times higher intensity at 532 nm (Fig. 1(c)). Unfortunately, such an approach would lead to increased OPF and reduced temporal pulse contrast eventually not suitable for high-field physics [26, 31]. Moreover, such a high pump intensity is above the BBO damage threshold for this pump pulse duration. High contrast amplification of sub-three-cycle, TW light pulses was previously demonstrated using 5 mm long type-I BBO at moderate pump intensities [13]. Moreover, investigations on TBP-NOPCPA are also of high interest for multiple-color-pumping. For instance, one can perform pumping it with the second-harmonic of Nd:YAG and with the second-harmonic of Yb:YAG (at 515 nm) in the same NOPCPA stage to enhance the signal spectrum. A proposed phase-matching for this case is shown in Fig. 1(a) and (d). For example, the important question has to be answered, if a signal spectrum consisting of disjunct spectral components amplified by different pump beams can actually be compressed. The compressibility of such a spectrum is shown in the present investigations.

3. Experimental setup

The experimental TBP-NOPCPA set up is shown in Fig. 2(b). We use a part of the setup of a sub-three-cycle, 16-TW NOPCPA laser system described in Ref. [13], which is used for high-field experiments such as high-order harmonic generation from overdense plasmas and electron acceleration [7, 4]. As seed source for the TBP-NOPCPA we use 2 μJ pulses with a spectrum ranging from 650 nm to 970 nm (see Fig. 4(a)) and a FWHM pulse duration of 22 ps at 10 Hz repetition rate. The seed is negatively chirped by using a reflection grism-pair and an acousto-optic modulator (Dazzler, Fastlite). The group delay between the spectral boundaries of the stretched seed is 42 ps. The pump pulses for the TBP-NOPCPA are generated with a commercial Nd:YAG laser amplifier (Ekspla), which provides up to 1 J pulses at 532 nm with a FWHM pulse duration of 78 ps at 10 Hz repetition rate. The seed and pump pulses are synchronized in an all-optical way. The TBP-NOPCPA stage consists of a 5 x 5 x 5 mm^3 type-I BBO crystal and is pumped with variable total pump energy up to 12 mJ in two beams. The pump beam is relay imaged from the pump laser output on the BBO crystal and split into two parts (pump

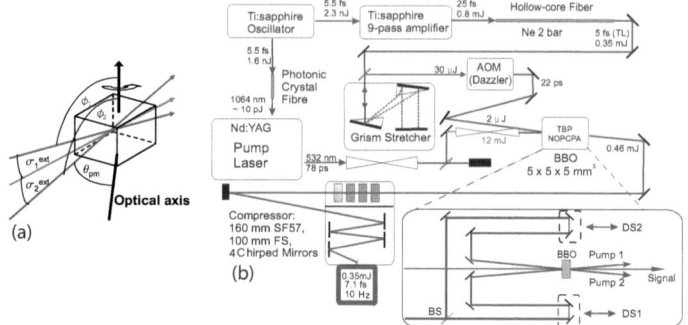

Fig. 2. (a) Optical schematic of the TBP-NOPCPA phase-matching geometry. (b) Layout of the TBP-NOPCPA experimental set up: BS - beamsplitter, DS - delay stage, AOM - acousto-optic modulator.

1 and pump 2) with a 50%:50% beamsplitter. Each pump beam has its own delay stage to achieve temporal overlap between the pump pulses and the seed pulse. The amplified signal beam is expanded from about 2 mm to 18 mm and compressed in bulk material consisting of 160 mm SF57 (Schott) and 100 mm fused silica. After the bulk compressor, the beam is down-collimated to about 6 mm and led into a vacuum compression chamber for final compression with 4 positive-dispersion chirped mirrors. The stretcher and compressor together have a calculated B-integral of 0.08 and the compressor has a throughput of 75% (including several silver mirrors).

The pulse duration is measured with a home-built all-reflective single-shot second-order autocorrelator, which is described in section 4.

4. All-reflective single-shot second-order intensity autocorrelator for few-cycle pulse characterization

Single-shot autocorrelation set ups have been demonstrated many years ago [37][38], as well as all-reflective multi-shot autocorrelators being able to measure pulse durations of only a few femtoseconds [39]. To our knowledge, up to now no all-reflective single-shot second-order intensity autocorrelator was realized and used to characterize light pulses with a duration of only a few cycles and a spectral bandwidth exceeding 300 nm. The following presented set up was specially designed to measure the duration and intensity distribution of the pulses generated in Ref. [13] and in the present experiment.

Figure 3 shows the set up of our home-built all-reflective single-shot second-order intensity autocorrelator. The main feature of this device is its minimized dispersion as well as the temporal resolution of about 197 attoseconds, which allows the characterization of few-cycle light pulses. A broad spectral range, from 675 nm up to 1000 nm, made the realization to a major challenge.

The set up consists of two irises at the entrance of the autocorrelator allowing an accurate alignment of the incoming beam. For splitting the beam in two identical replicas, geometric beamsplitting is used, which is done by a D-shaped mirror. This method is used to minimize dispersion in the optical path of the fundamental pulses. Due to the broad spectrum of few-

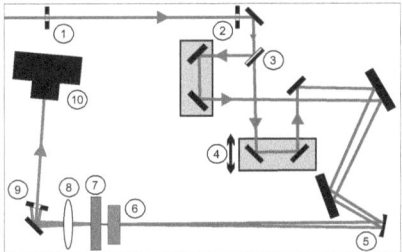

Fig. 3. Setup of the single-shot all-reflective second-order intensity autocorrelator with a temporal resolution of about 197 as/px and an observation window of 565 fs. 1 & 2: entrance iris, 3: geometric beamsplitter, 4: delay stage with micrometer screw, 5: cylindric mirror, 6: SHG crystal, 7: color-glass filter, 8: achromatic imaging lens, 9: small aperture iris, 10: CCD detector (Larry 3000 USB)

cycle pulses even thin dispersive elements, like plate beam-splitters, would cause significant distortion of the original pulse. Pellicle beam-splitters are not chosen because of their sensitivity to vibrations. The upper part of the beam propagates in forward direction into the delay stage, whereas the lower part of the beam is deflected by 90°. In single-shot devices, the delay stage is only used for finding the zero-delay at the intersection point in the nonlinear optical crystal and for calibration. Unlike in a multi-shot autocorrelator the two pulse replicas are overlapped in the nonlinear optical crystal under a horizontal noncollinear angle of $\sigma_{nc} = 2.34°$. A cylindric mirror with a focal length of 300 mm placed in front of the crystal generates line foci in the horizontal plane with a vertical diameter of about 51 μm. This method drastically reduces the required pulse energy to 10 μJ. With a phase-matching angle $\Theta_{pm} = 42°$ and a thickness of the BBO crystal of 5 μm, type-I phase-matching for the whole spectral bandwidth of the fundamental is provided, which allows phase matching starting at 465 nm and exceeding 1100 nm. The wavelength which has perfect phase matching was chosen to be 583 nm. The deviation from the laser central wavelength, located at 790 nm, allows a better phase matching in the short wavelength region while maintaining acceptable phase matching at long wavelengths.

For alignment purpose, the fundamental light is blocked behind the BBO crystal with a BG37 color glass filter. The second harmonic signal is then imaged by an achromatic lens with a focal length of 40 mm and four-time magnified from the crystal plane to the detector. The detector is a linear CCD array (Larry 3000 USB, Ames Photonics) with 3000 pixels in a horizontal row and a width of 7 μm each. In the focal plane of the imaging lens, the two second harmonic signals, propagating in direction of the fundamental beams, are blocked by a small aperture to prevent deteriorations of the autocorrelation signal. The isolated second harmonic photons at the bisector of the noncollinear angle σ_{nc}, carrying the intensity autocorrelation function in their transversal spatial intensity profile, are recorded by the detector mentioned above. An additional BG37 color glass filter reduces fundamental stray light reaching the detector.

The temporal resolution and observation window, being the two important properties, are defined by the maximum width of the signal, which is limited by the aperture of the BBO crystal d = 5 mm, and the noncollinear angle σ_{nc}. Using basic geometry, the temporal observation window τ_{obs} can be expressed by

$$\tau_{obs} = \frac{n \cdot d \cdot \sin(\sigma_{nc}/2)}{c} \qquad (13)$$

where c denotes the velocity of light in vacuum and n the ordinary refractive index of BBO at 790 nm. For $\sigma_{nc} = 2.34°$ used in the setup, the obtained observation window has a size of τ_{obs}=565 fs (Eq. (13)). The CCD array with a width of 21 mm, containing 3000 pixels, records a four-times magnified image of the signal with an extension of 20 mm. Hence, the achieved temporal resolution r_{AC} results in the observation window divided by the number of pixels, covered by the signal, and has a value of r_{AC} = 197 $\frac{as}{px}$ for $\sigma_{nc} = 2.34°$.

To estimate the absolute error of the pulse duration measured with the autocorrelator, the single errors have to be investigated. The pulse duration observed by autocorrelation consists of three variables

$$\tau_{pulse} = \frac{\tau_{AC} \cdot r_{AC}}{K} \qquad (14)$$

where K denotes the deconvolution factor, τ_{AC} the FWHM of the autocorrelation trace in pixels and r_{AC} the temporal resolution of 197 $\frac{as}{px}$ as mentioned above. The FWHM pulse duration (Eq. (14)) is obtained by dividing the temporal FWHM of the autocorrelation trace ($\tau_{AC} \cdot r_{AC}$) by the deconvolution factor K. This value is unique and strongly dependent on the pulse shape and hence strongly influences the measurement error. We calculated the deconvolution factor for the Fourier transformed pulses of several different spectra of our TW laser system, ending up with a value of K=1.35± (1 ·10^{-3}). The change in the converted spectrum by the phase-matching curve of the BBO crystal has also been considered in this calculation.

Fortunately, in single-shot SHG autocorrelation measurement no geometrical distortions occur, since the generated second harmonic signal propagates along the bisector of angle formed by the input beams. Consequently there are no geometrical effects, decreasing the accuracy of the measurement [40].

An error originating from the calibration Δr_{AC} was calculated to be 4 $\frac{as}{px}$, while an error of τ_{AC} could be reduced by interpolation to $\Delta \tau_{AC}$ = 0.1 px.

Gaussian error propagation consequently leads to a total error of 0.12 fs for a fourier limited pulse. The deviation of the measured pulse from the Fourier limited pulse duration causes some deviation of the deconvolution factor K. Consequently the measurement error is influenced by this deviation which is hard to estimate, but as pulse duration gets close to the transform limit this error is also supposed to become very small.

For measuring the pulses of our present set up, the beam was coupled out of the compressor chamber through a 3 mm thick fused silica window. In addition to the window and some air, further dispersion of about 40 fs^2 is caused by reflections from seven protection-coated silver mirrors and the 5 μm thick BBO crystal inside the autocorrelator. Of course this dispersion is compensated by the AOM when measuring the optimized pulse duration.

Altogether the presented all-reflective single-shot second-order intensity autocorrelator was specially designed to characterize few-cycle pulses. With a temporal observation window of 565 fs and a temporal resolution of about 197 as, it is well suited to measure duration and intensity distribution of few-cycle pulses at realtime with a repetition rate of 10 Hz given by the laser system.

5. Results and discussion

The seed is amplified in the TBP-NOPCPA geometry shown in Fig. 2(a) and a signal energy up to 0.46 mJ is achieved. The broadest measured signal pulse spectrum is shown in Fig. 4(b). The signal pulse spectrum ranges from 675 nm to 970 nm leading to a Fourier limit of 6.7 fs. The TBP setup was modified to operate it also as OBP-NOPCPA with the same total pump energy by replacing the beamsplitter with a mirror. A typical signal spectrum in the OBP-NOPCPA case is shown in Fig. 4(a). This spectrum ranges from around 700 nm up to 970 nm leading to a Fourier limit of 8.5 fs. The spectral limits in Fig. 4(b) at low wavelength (675 nm and 700

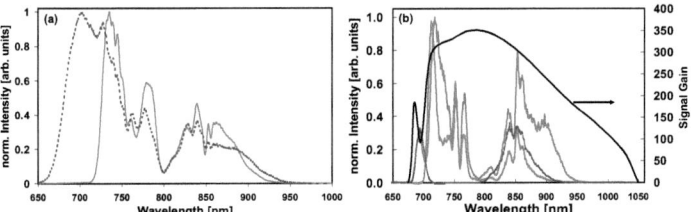

Fig. 4. (a) Unamplified seed spectrum (blue dotted) and amplified signal spectrum (red) in case of OBP-NOPCPA by phase-matching almost the same bandwidth as in Ref. [13] with $\sigma = 2.23°$ and $\theta_{pm} = 23.62°$. (b) Amplified signal spectrum using the TBP-NOPCPA scheme (red) with $\sigma_1 = 2.22°, \sigma_2 = -2.16°$ and $\theta_{pm} = 23.62°$, the corresponding simulated gain (black), amplified signal spectrum using only pump1 (green) and using only pump2 (blue dotted).

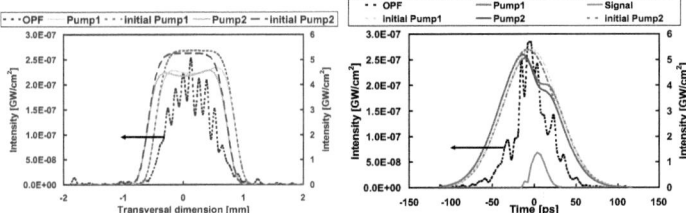

Fig. 5. Initial pump pulses and pump, signal and OPF after amplification in the time and space domain.

nm) are due to the different phase-matching geometry (see Fig. 1) and at the long wavelength are due to a lack of seed components above 970 nm, although amplification is theoretically possible up to 1050 nm (Fig. 4(b)). The dip in the signal spectra at around 800 nm results from the corresponding dip in the unamplified seed spectrum, which is a consequence of the self-phase modulation taking place in the hollow-core fiber.

Figure 4(b) shows the spectral parametric signal intensity gain obtained with the numerical split-step simulation described in section 2. The simulated amplified signal energy content is 0.49 mJ distributed over a spectral bandwidth from 675 nm to 970 nm, which matches the experimental values. Our simulations show an OPF peak intensity of $2.8 \cdot 10^{-7}$ GW/cm^2 and a signal peak intensity of 1.4 GW/cm^2 after amplification. This leads to an expected pulse contrast ratio of $6 \cdot 10^{-11} - 10^{-10}$ on the picosecond timescale after compression, assuming the beam transport efficiency and compressor transmission to be the same for signal and OPF. The OPF after amplification in space and time relative to the pump pulses and the signal pulse at the end of the BBO crystal is shown in Fig. 5. OPF does only occur within the temporal and spatial window of the pump pulse and shows strong amplitude modulation in the time and space domain.

Additionally, we measure the signal energy as function of pump intensity and calculate the energy pump-to-signal conversion efficiency. The results are shown in Fig. 6(a). For this rea-

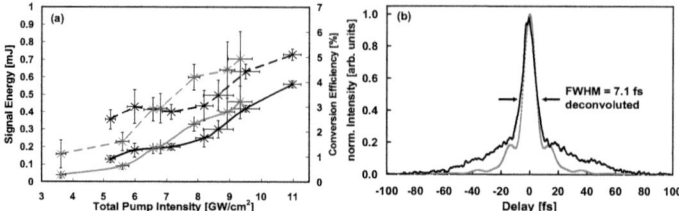

Fig. 6. (a) The dotted curves show the TBP-NOPCPA pump-to-signal conversion efficiency for the case of the red signal spectrum shown in 4(b) (green) and for the case of the signal spectrum similar to the red curve shown in 4(a)(black) with $\sigma_1 = -\sigma_2 = 2.23°$ and $\theta_{pm} = 23.62°$. The solid curves show the corresponding signal energy as function of total pump intensity. (b) Measured second-order single-shot autocorrelation (solid black curve) and calculated autocorrelation trace (red dotted curve)

son, we are able to investigate, if the phase-matching ($\sigma_1 = -\sigma_2 = 2.23°$ and $\theta_{pm} = 23.62°$) for the TBP-NOPCPA geometry aiming for a spectrum similar to one also achievable with OBP-NOPCPA (the case of Fig. 4(a)) is comparable to the conversion efficiency of the TBP-NOPCPA phase-matching geometry in case of Fig. 4(b) with broader amplified bandwidth. As result, the conversion efficiency is comparable within the error bars for both cases. Moreover, we measure the pulse duration of the amplified and compressed broadband signal pulse via an all-reflective second-order single-shot autocorrelation. The results are shown in Fig. 6(b) and reveal a FWHM pulse duration of 7.1 fs, whereas the Fourier-limited FWHM pulse duration is typically 6.7 fs. Consequently, compression is achieved to within 6% of the FWHM Fourier-limit. This important fact reveals the compressibility of the whole amplified signal bandwidth, although this signal spectrum is composed of two individual NOPCPA interactions taking place in the same nonlinear crystal. A deconvolution factor of 1.4 is calculated from the spectrum. TBP-NOPCPA also makes an additional spectral shaping possible compared to OBP-NOPCPA; i.e. shaping of the steep edge at 700 nm in the amplified signal spectrum in Fig. 4(a). This can potentially prevent pulse satellites from occuring after compression if a saturated TBP-NOPCPA stage is used, which would be present due to a sharp spectral edge even in case of compression close to the Fourier-limit [11, 12, 13]. In this case, a better pulse contrast is achievable due to an improvement of the foot of the compressed pulse, using a saturated amplifier stage.

Furthermore, there have been concerns, that a transient grating generated by the intersecting pump beams could potentially degrade the signal beam quality, since the amplification strongly depends on the pump intensity. For this reason, we investigate if there is an intensty grating (due to interference) present inside the BBO crystal and whether this grating negatively affects the signal beam quality. For this purpose, we image the BBO crystal plane with an achromatic lens onto a CCD camera with a pixel size of 6.7 x 6.7μm^2. We magnify the object with a calibrated factor of 11 and verify, that crystal defects in the crystal are sharp on the CCD image in any case.

Firstly, we imaged only the intersection of the two pump beams and recorded the magnified crystal plane as a function of relative polarization orientation of the two beams. In this case, the seed was blocked before the crystal. The images and their lineouts are shown in Fig. 7(c) and (a), where 0° denotes parallel polarization orientation of the two pump beams and 90° corresponds to orthogonal relative polarization orientation. We found, that if the polarization of the two pump beams is parallel to each other and thus allow for interference taking place,

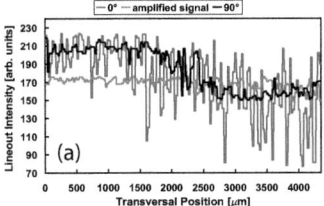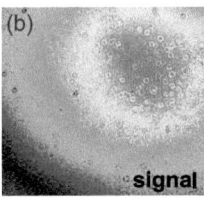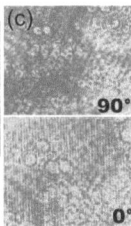

Fig. 7. Image of the BBO crystal plane magnified by factor 11 on a CCD camera: (a) Image line outs of the crystal plane. (b) Image of the amplified signal beam profile. (c) Image of the intersecting pump beam profiles in the crystal plane with varying relative polarization orientation: (0° denotes parallel, 90° denotes orthogonal polarization of the two pump beams).

there is clearly a pump intensity modulation present (see Fig. 7) in the crystal plane with a period of $\sim 7\mu m$, taking the magnification factor into account. This modulation vanishes in case of orthogonal polarization between the two pump beams, because then there is no interference possible. Consequently, this optical set up is able to resolve an intensity grating due to interference inside the BBO crystal.

Secondly, we unblocked the seed and measured the magnified image of the amplified signal in the crystal plane, while the pump beams are blocked after passing through the crystal. Figure 7(b) and (a) show, that we observed no signal intensity modulation due to the pump interference in the beam profile and the corresponding lineout. This can be explained as follows:

The transient grating due to the interference between the two pump beams inside the BBO crystal evolves on the angle bisector in between the pump beams. Since the two noncollinear angles are not the same, the signal beam traverses this bisector leading to a transversal shift with respect to the bisector of half a grating period over the crystal length. Consequently, this smearing effect can prevent the pump grating from degrading the signal beam quality. Apart from general noncollinear smearing, other effects like the Poynting vector walk-off can also contribute to this effective smearing.

6. Conclusion

We outlined a new concept (TBP-NOPCPA) for the generation of few-cycle, high-contrast, multi-terawatt light pulses. Moreover, the present experiment demonstrated compressibility of an amplified signal pulse, whose spectral components result from amplification by two separate pump beams in the same nonlinear crystal. In this context, we experimentally showed amplification and compressibility of the signal between 675 nm and 970 nm leading to 7.1-fs, 0.35-mJ light pulses after compression to within 6% of the FWHM Fourier-limit. Technical details of few-cycle pulse characterization are outlined. We investigated geometry, phase-mismatch, signal gain bandwidth, pump-to-signal conversion efficiency and signal beam quality for this new broadband ϕ-plane-pumped TBP-NOPCPA scheme. Additionally, we performed numerical split-step simulations of TBP-NOPCPA to obtain the parametric gain curve and to study the level of OPF. The temporal pulse contrast ratio was simulated to be around 10^{-10} on ps timescales after amplification and compression. It was shown that the accumulated OPA phase partially compensates for wave-vector mismatch to maintain maximum gain and leads to extended broadband amplification. Summing up, the present investigations revealed, that TBP-NOPCPA is a promising parametric scheme without a decrease in amplified signal compressibility, amplified signal beam quality and conversion efficiency, compared to conventional one-beam-pumping NOPCPA. The presented theoretical considerations and numerical simulations showed, that amplification is possible between 675 nm and 1050 nm in a 5 mm long BBO crystal. Consequently, high-contrast, sub-7-fs, multi-TW pulses are achievable with a 5 mm long type-I BBO crystal, pumped at 532 nm, in case of TBP-NOPCPA.

Acknowledgements

This work was supported by Deutsche Forschungsgemeinschaft (contract TR18), the association EURATOM-Max-Planck-Institut für Plasmaphysik and by the Cluster of Excellence Munich center for Advanved Photonics (MAP). D. H., who mainly worked on experimental and theoretical TBP-NOPCPA, is also grateful to Studienstiftung des deutschen Volkes. R. T., who mainly worked on pulse diagnostics and the seed source, is thankful to Deutsche Forschungsgemeinschaft (contract TR18).

Appendix B5

Approaching the full octave: Noncollinear optical parametric chirped pulse amplification with two-color pumping

D. Herrmann, C. Homann, R. Tautz, M. Scharrer, P. St.J. Russell, F. Krausz, L. Veisz, and E. Riedle

Reprinted with permission from

Optics Express 18, 18752-18762 (2010).

DOI: 10.1364/OE.18.018752

http://www.opticsinfobase.org/oe/abstract.cfm?URI=oe-18-18-18752

© 2010 The Optical Society of America

Approaching the full octave: Noncollinear optical parametric chirped pulse amplification with two-color pumping

D. Herrmann,[1,2,*] C. Homann,[2] R. Tautz,[1,3] M. Scharrer,[4] P. St.J. Russell,[4] F. Krausz,[1,5] L. Veisz,[1,6] and E. Riedle[2]

[1] *Max-Planck-Institut für Quantenoptik, Hans-Kopfermann-Str. 1, 85748 Garching, Germany*
[2] *LS für BioMolekulare Optik, LMU München, Oettingenstr. 67, 80538 München, Germany*
[3] *present address: LS für Photonik und Optoelektronik, LMU München, Amalienstr. 54, 80799 München, Germany*
[4] *Max-Planck-Institut für die Physik des Lichts, Günther-Scharowsky-Str. 1/ Bau 24, 91058 Erlangen, Germany*
[5] *LS für Laserphysik, LMU München, Am Coulombwall 1, 85748 Garching, Germany*
[6] *laszlo.veisz@mpq.mpg.de*
[*] *d.herrmann@physik.uni-muenchen.de*

Abstract: We present a new method to broaden the amplification range in optical parametric amplification toward the bandwidth needed for single cycle femtosecond pulses. Two-color pumping of independent stages is used to sequentially amplify the long and short wavelength parts of the ultrabroadband seed pulses. The concept is tested in two related experiments. With multi-mJ pumping pulses with a nearly octave spanning spectrum and an uncompressed energy of 3 mJ are generated at low repetition rate. The spectral phase varies slowly and continuously in the overlap region as shown with 100 kHz repetition rate. This should allow the compression to the Fourier limit of below 5 fs in the high energy system.

©2010 Optical Society of America

OCIS codes: (190.7110) Ultrafast nonlinear optics; (190.4970) Parametric oscillators and amplifiers; (190.2620) Harmonic generation and mixing; (320.5520) Pulse compression; (190.4975) Parametric processes; (260.7120) Ultrafast phenomena

References and links

1. F. Krausz, and M. Ivanov, "Attosecond physics," Rev. Mod. Phys. **81**(1), 163–234 (2009).
2. G. D. Tsakiris, K. Eidmann, J. Meyer-ter-Vehn, and F. Krausz, "Route to intense single attosecond pulses," N. J. Phys. **8**(1), 19 (2006).
3. T. Tajima, "Laser acceleration and its future," Proc. Jpn. Acad. Ser. B **86**(3), 147–157 (2010).
4. M. Nisoli, S. De Silvestri, and O. Svelto, "Generation of high energy 10 fs pulses by a new pulse compression technique," Appl. Phys. Lett. **68**(20), 2793–2795 (1996).
5. C. P. Hauri, W. Kornelis, F. W. Helbing, A. Heinrich, A. Couairon, A. Mysyrowicz, J. Biegert, and U. Keller, "Generation of intense, carrier-envelope phase-locked few-cycle laser pulses through filamentation," Appl. Phys. B **79**(6), 673–677 (2004).
6. G. M. Gale, M. Cavallari, T. J. Driscoll, and F. Hache, "Sub-20-fs tunable pulses in the visible from an 82-MHz optical parametric oscillator," Opt. Lett. **20**(14), 1562–1564 (1995).
7. T. Wilhelm, J. Piel, and E. Riedle, "Sub-20-fs pulses tunable across the visible from a blue-pumped single-pass noncollinear parametric converter," Opt. Lett. **22**(19), 1494–1496 (1997).
8. A. Baltuška, T. Fuji, and T. Kobayashi, "Visible pulse compression to 4 fs by optical parametric amplification and programmable dispersion control," Opt. Lett. **27**(5), 306–308 (2002).
9. S. Adachi, N. Ishii, T. Kanai, A. Kosuge, J. Itatani, Y. Kobayashi, D. Yoshitomi, K. Torizuka, and S. Watanabe, "5-fs, Multi-mJ, CEP-locked parametric chirped-pulse amplifier pumped by a 450-nm source at 1 kHz," Opt. Express **16**(19), 14341–14352 (2008).
10. I. N. Ross, P. Matousek, M. Towrie, A. J. Langley, and J. L. Collier, "The prospects for ultrashort pulse duration and ultrahigh intensity using optical parametric chirped pulse amplification," Opt. Commun. **144**(1-3), 125–133 (1997).
11. S. Witte, R. T. Zinkstok, A. L. Wolf, W. Hogervorst, W. Ubachs, and K. S. E. Eikema, "A source of 2 terawatt, 2.7 cycle laser pulses based on noncollinear optical parametric chirped pulse amplification," Opt. Express **14**(18), 8168–8177 (2006).
12. D. Herrmann, L. Veisz, R. Tautz, F. Tavella, K. Schmid, V. Pervak, and F. Krausz, "Generation of sub-three-cycle, 16 TW light pulses by using noncollinear optical parametric chirped-pulse amplification," Opt. Lett. **34**(16), 2459–2461 (2009).

13. T. S. Sosnowski, P. B. Stephens, and T. B. Norris, "Production of 30-fs pulses tunable throughout the visible spectral region by a new technique in optical parametric amplification," Opt. Lett. **21**(2), 140–142 (1996).
14. E. Zeromskis, A. Dubietis, G. Tamosauskas, and A. Piskarskas, "Gain bandwidth broadening of the continuum-seeded optical parametric amplifier by use of two pump beams," Opt. Commun. **203**(3-6), 435–440 (2002).
15. D. Herrmann, R. Tautz, F. Tavella, F. Krausz, and L. Veisz, "Investigation of two-beam-pumped noncollinear optical parametric chirped-pulse amplification for the generation of few-cycle light pulses," Opt. Express **18**(5), 4170–4183 (2010).
16. G. Tamošauskas, A. Dubietis, G. Valiulis, and A. Piskarskas, "Optical parametric amplifier pumped by two mutually incoherent laser beams," Appl. Phys. B **91**(2), 305–307 (2008).
17. C. Schriever, S. Lochbrunner, P. Krok, and E. Riedle, "Tunable pulses from below 300 to 970 nm with durations down to 14 fs based on a 2 MHz ytterbium-doped fiber system," Opt. Lett. **33**(2), 192–194 (2008).
18. C. Homann, C. Schriever, P. Baum, and E. Riedle, "Octave wide tunable UV-pumped NOPA: pulses down to 20 fs at 0.5 MHz repetition rate," Opt. Express **16**(8), 5746–5756 (2008).
19. M. Bradler, P. Baum, and E. Riedle, "Femtosecond continuum generation in bulk laser host materials with sub-µJ pump pulses," Appl. Phys. B **97**(3), 561–574 (2009).
20. G. Cerullo, M. Nisoli, S. Stagira, and S. De Silvestri, "Sub-8-fs pulses from an ultrabroadband optical parametric amplifier in the visible," Opt. Lett. **23**(16), 1283–1285 (1998).
21. I. Z. Kozma, P. Baum, U. Schmidhammer, S. Lochbrunner, and E. Riedle, "Compact autocorrelator for the online measurement of tunable 10 femtosecond pulses," Rev. Sci. Instrum. **75**(7), 2323–2327 (2004).
22. P. Baum, S. Lochbrunner, and E. Riedle, "Zero-additional-phase SPIDER: full characterization of visible and sub-20-fs ultraviolet pulses," Opt. Lett. **29**(2), 210–212 (2004).
23. S. Witte, R. T. Zinkstok, W. Hogervorst, and K. S. E. Eikema, "Numerical simulations for performance optimization of a few-cycle terawatt NOPCPA system," Appl. Phys. B **87**(4), 677–684 (2007).
24. A. L. Cavalieri, E. Goulielmakis, B. Horvath, W. Helml, M. Schultze, M. Fiess, V. Pervak, L. Veisz, V. S. Yakovlev, M. Uiberacker, A. Apolonski, F. Krausz, and R. Kienberger, "Intense 1.5-cycle near infrared laser waveforms and their use for the generation of ultra-broadband soft-x-ray harmonic continua," N. J. Phys. **9**(7), 242 (2007).
25. J. Park, J. H. Lee, and C. H. Nam, "Generation of 1.5 cycle 0.3 TW laser pulses using a hollow-fiber pulse compressor," Opt. Lett. **34**(15), 2342–2344 (2009).
26. A. Baltuška and T. Kobayashi, "Adaptive shaping of two-cycle visible pulses using a flexible mirror," Appl. Phys. B **75**(4-5), 427–443 (2002).
27. I. N. Ross, P. Matousek, G. H. C. New, and K. Osvay, "Analysis and optimization of optical parametric chirped pulse amplification," J. Opt. Soc. Am. B **19**(12), 2945–2956 (2002).
28. A. Renault, D. Z. Kandula, S. Witte, A. L. Wolf, R. Th. Zinkstok, W. Hogervorst, and K. S. E. Eikema, "Phase stability of terawatt-class ultrabroadband parametric amplification," Opt. Lett. **32**(16), 2363–2365 (2007).
29. G. Rodriguez, and A. J. Taylor, "Measurement of cross-phase modulation in optical materials through the direct measurement of the optical phase change," Opt. Lett. **23**(11), 858–860 (1998).
30. G. Krauss, S. Lohss, T. Hanke, A. Sell, S. Eggert, R. Huber, and A. Leitenstorfer, "Synthesis of a single cycle of light with compact erbium-doped fibre technology," Nat. Photonics **4**(1), 33–36 (2010).

1. Introduction

High energy ultrafast light pulses are unique tools for applications ranging from the most fundamental science to medical applications. Quasi-single cycle near infrared pulses can generate single attosecond pulses or even single cycle attosecond pulses in the XUV that allow the measurement of the fastest known processes [1,2]. Even with slightly longer pump pulses highly attractive electron acceleration has been shown that offers a tabletop alternative to large scale facilities used in clinical environments [3].

With known laser materials used in multi-stage optically pumped chirped pulse amplifiers the attainable spectral width is limited by either the intrinsic gain bandwidth or even more severely the spectral gain narrowing in going from nJ broadband seed light to the desired Joule output levels. To overcome these limitations, frequency broadening in either gas filled hollow capillary fibers [4] or in filaments [5] is widely used. So far the use is, however, limited to at most a few mJ.

An attractive alternative for the direct generation of high energy extremely broadband pulses is optical parametric amplification. With a noncollinear geometry and pumping by a frequency-doubled femtosecond Ti:sapphire laser, sub-20 fs visible pulses have indeed been generated at 82 MHz repetition rate [6] and multi-µJ pulses at kHz rates [7]. The concept of the noncollinearly phase-matched optical parametric amplifier (NOPA) was optimized to the generation of 4 fs pulses [8].

Pumping by femtosecond Ti:sapphire pulses limits the attainable output energy, even though extremely short pulses with peak powers of 0.5 TW have been demonstrated [9]. Therefore the use of pico- or even nanosecond pump pulses was suggested in combination with a chirped pulse strategy (OPCPA) [10]. All known and technically available pump lasers with ps pulse duration operate around 1050 nm due to the laser active materials used, e.g. Nd^+ or Yb^+. The use of such pump lasers has already led to 2 TW pulses [11] and lately sub-8 fs pulses centered at 805 nm with more than 130 mJ (16 TW) compressed energy [12]. A further shortening of the pulses in this spectral region becomes increasingly difficult due to the phase-matching bandwidth of the BBO amplifier medium with one-color pumping.

Already 1996 it was shown that multiple amplification stages with slight spectral detuning of the individual amplification range could lead to significant shortening of the output in a 100 kHz OPA [13]. Pumping of a single stage with multiple pump beams allowed the demonstration of pulse shortening from 98 to 61 fs [14] and lately from 8 to 7 fs with an improved temporal structure [15]. The use of different pump wavelengths was even suggested, yet no spectral nor temporal characterization of the output has been provided [16].

For a MHz Yb-based femtosecond pump system it was shown that both the green 2ω light and the UV 3ω light can be used to pump a NOPA [17,18]. Both configurations easily yield sub-20 fs pulse durations even without optimized compression. Interestingly, the 2ω pumping produces these pulses in the red and the 3ω pumping in the green and yellow part of the spectrum. We therefore suggest the simultaneous use of green and UV pumped stages together with a chirped pulse scheme and high energy pumping. This should allow the generation of pulses with unprecedented shortness and pulse energy.

In this communication we report on two pilot experiments that are aimed on testing this hypothesis. With tens of mJ pumping the generation of nearly octave-spanning few-mJ pulses is achieved. For this effort, techniques have been developed to allow efficient frequency conversion of the 78 ps fundamental pump pulses at 1064 nm. In a second experiment at 100 kHz repetition rate the compressibility and spectral phase behavior of the composite pulses is investigated.

2. NOPCPA on the mJ-level approaching the octave

Figure 1 shows the layout of the experimental two-color-pumped NOPCPA setup. We have extended an existing system (denoted as Light-Wave-Synthesizer-20, Ref. [12]), which generates sub-8 fs, 130-mJ pulses at 805 nm central wavelength and 10 Hz repetition rate.

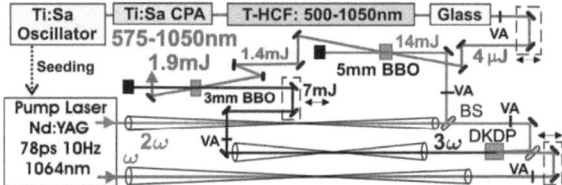

Fig. 1. Layout of the mJ-level two-color-pumped NOPCPA setup, which consists of two cascaded stages employing type-I phase-matching in BBO (T-HCF – tapered hollow-core capillary fiber, VA – variable attenuator, BS – beam splitter). The pump fundamental (1064 nm, ω) and its second- (532 nm, 2ω) and third-harmonic (355 nm, 3ω) are relay-imaged with vacuum telescopes onto the nonlinear optical crystals BBO and DKDP.

The present system consists of a Ti:sapphire master oscillator from which the broadband seed and the pump pulses for the two-stage NOPCPA chain are derived. 60% of the oscillator output is fed into a Ti:sapphire 9-pass chirped-pulse amplifier (Femtopower Compact pro, Femtolasers GmbH), delivering 25-fs, 800-µJ pulses at 1 kHz. After amplification, the pulses are spectrally broadened in a 1 m long tapered hollow-core capillary fiber (T-HCF) filled with neon gas at 2 bar (1500 Torr) absolute pressure, which leads to a seed spectrum ranging from 500 to 1050 nm (30 dB; see Fig. 2). The larger diameter of 500 µm in the first 10 cm together with the small diameter of 200 µm in the main part helps broaden the spectrum by up to

50 nm in the red wing. These pulses are then stretched in time by a variable amount of glass, leading to an adjustable group delay (GD) of the seed spectral boundaries. The seed pulse energy is attenuated to 4 µJ to be comparable with the seed used in Ref. [12].

The remaining 40% of the oscillator output is used to seed the Nd:YAG pump laser (EKSPLA, UAB), providing all-optical synchronization. This pump laser delivers two beams of 78-ps (FWHM), up to 1-J pulses at 1064 nm and 10 Hz repetition rate. Pulses at 532 nm are generated via type-II second-harmonic generation (SHG) of two fundamental 1064 nm beams in a 10 mm long DKDP crystal. The remaining fundamental and a part of the SH are relay-imaged with vacuum telescopes onto a 10 mm long DKDP crystal for type-II sum-frequency generation (SFG). Usually, 6 mJ at 532 nm and 14 mJ at 1064 nm are used to generate 8 mJ pulses of the third-harmonic (TH) at 355 nm. The beam diameters are 5 mm each. A high energy conversion efficiency of 40% and a quantum efficiency of 90% with respect to the SH is observed. The SH and TH pump pulse durations are also approximately 78 ps due to the strong saturation in the conversion.

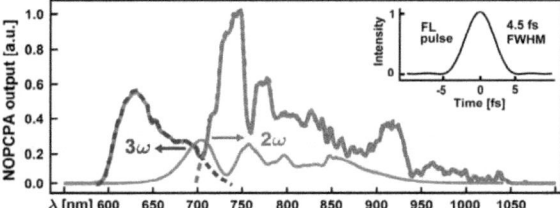

Fig. 2. The spectral energy density of the signal pulse amplified by the second-harmonic and the third-harmonic of the Nd:YAG pump laser (575-1050 nm, red solid curve) allows for a Fourier-limit of 4.5 fs (inset). This spectrum is composed of the spectral region amplified only by the second-harmonic (700-1050 nm, green dotted curve shown as guide to the eye) and the third-harmonic alone (575-740 nm, blue dotted curve). A typical output spectrum of the T-HCF (unamplified seed, not to scale) is shown as gray solid curve.

The first NOPCPA stage (calculated phase-matching angle $\theta = 23.62°$, internal noncollinear angle $\alpha = 2.23°$) consists of a 5 mm long type-I BBO crystal with a slight wedge to avoid adverse effects of internal reflections [9]. It is pumped by 14 mJ of the SH with a 2 mm diameter of the sixth-order super-Gaussian beam and a peak intensity of about 10 GW/cm^2. The second NOPCPA stage ($\theta = 34.58°$, $\alpha = 3.40°$) consists of a 3 mm long type-I BBO crystal and is pumped by 7 mJ of the TH with 1.5 mm diameter and a peak intensity of close to 10 GW/cm^2. Both stages are operated in tangential phase-matching geometry, both pump beams being relay-imaged onto the crystals and being slightly smaller than the seed beam so as to achieve a good spatial signal beam profile. With a group delay of 37 ps for the full seed bandwidth, the seed is amplified from 700 to 1050 nm to an energy of 1.4 mJ in the first stage and further amplified from 575 to 740 nm in the second stage. The particular seed delay was chosen to ensure that the full spectral range of the seed lies within the pump pulse duration. The very high small signal gain of OPCPA allows utilization of the exponentially decreasing seed light at the spectral edges. Overall, an output energy of 1.9 mJ and a spectrum spanning from 575 to 1050 nm is achieved. This nearly octave-wide spectrum supports a Fourier limited pulse of 4.5 fs (see inset in Fig. 2). The results from the seed GD variation are discussed in section 4.1. The positively chirped pulses do not allow the use of the available pulse compressor and hence restrict us to a spectral characterization of the pulses.

3. Proof-of-principle compressibility with high-repetition rate NOPA on the µJ-level

Figure 3 shows the layout for the µJ-level, high-repetition rate NOPA setup. As primary pump laser we use a commercial diode-pumped Yb:KYW disc laser system (JenLas® D2.fs,

Jenoptik AG), delivering 300-fs, 40-µJ pulses at a center wavelength of 1025 nm and a repetition rate of 100 kHz. For generating the seed we split off approximately 1.5 µJ of energy and focus it (all focal lengths are given in Fig. 3) onto a 2 mm thick YAG plate, where a supercontinuum ranging from 470 nm to above 1 µm results [19]. The main part of the pump pulses is focused towards two BBO crystals, where type-I SHG and subsequent type-II SFG are performed in a simple collinear arrangement [18]. In this way we obtain pulses with an energy of 13.5 µJ at the SH and 7.5 µJ at the TH, which are separated by dichroic mirrors and independently collimated with fused silica lenses. The supercontinuum is collimated with a thin fused silica lens and fed into the first NOPA stage ($\theta = 24.0°$, $\alpha = 2.3°$), which consists of a 3 mm long type-I BBO crystal. The first NOPA stage is pumped by the SH, whose energy can be adjusted by a combination of half-wave-plate and polarizer.

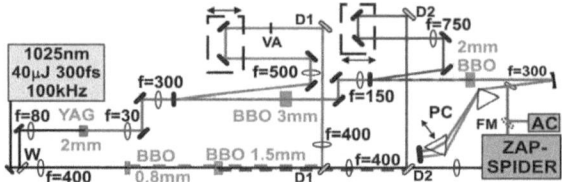

Fig. 3. Layout of the two-color-pumped NOPA setup, which consists of two cascaded stages employing type-I phase-matching in BBO (W - fused silica window, D - dichroic mirror, VA - variable attenuator, PC - prism compressor, FM - flipper mirror, AC - 2nd-order intensity autocorrelator). All focal lengths f are given in mm.

To achieve the necessary pump intensities for a high gain in the NOPA process, focusing of the pump beams and consequently focusing of the seed beams towards the nonlinear crystals is inevitable with this low-energy system, in contrast to the system described above. To achieve a good beam profile with a homogenous spectral distribution after both amplification stages, we find that proper matching of the divergence of pump and seed in both stages is crucial. For this reason we measured the divergence of all relevant beams around the position of the nonlinear crystals with a CCD camera and carefully selected and determined the focal lengths and positions of the respective focusing lenses. The amplified signal of the first stage was then 1:1 relay-imaged onto the BBO crystal (2 mm, type-I) of the second NOPA stage ($\theta = 35.5°$, $\alpha = 2.7°$), which is pumped by the TH.

In each stage special care has to be taken to spatially overlap the amplified signal beam with the respective seed beam, which is not ensured automatically. We have observed, that only in case of the geometry with the pump polarization oriented in the plane defined by seed and pump, spatial overlap of amplified signal and seed can be achieved. Additionally, the BBO crystal had to be set up to Poynting-vector walk-off configuration. We conclude, that the birefringent nature of the BBO crystal and the corresponding pump walk-off can compensate the shift of the amplified signal with respect to the seed direction. In the original design of the NOPA the noncollinear seed pump interaction is arranged in a vertical plane, the polarization of the seed is horizontal and that of the pump vertical [6,7]. Many setups use a horizontal geometry for practical reasons. If the seed is polarized vertical and the pump horizontal in this arrangement, no physical difference originates [8,12,20]. For the horizontal seed polarization and the vertical pump polarization used by us, we explicitly compared the vertical and horizontal pump seed interaction geometry. We found that only for the vertical beam geometry, a proper overlap of seed and signal could be achieved. Slight deviations from the optimum noncollinear and/or phase-matching angle had to be accepted.

We amplify spectral regions from 690 to 830 nm with the second harmonic and from 630 to 715 nm with the third harmonic (Fig. 4(a)). Due to the relatively short pump pulses (estimated to be 220 fs for the SH and 180 fs for the TH) in comparison to the chirped supercontinuum seed (close to 1 ps), the amplified bandwidth is determined by the temporal overlap of pump and seed. and not the phase-matching bandwidth of the crystals. Variation of only the seed pump delay results in a shift of the spectrum. With a seed energy of 2 nJ, typical output energies are 1 µJ for the first stage when pumped with 9.3 µJ (200 µm diameter, peak intensity of 250 GW/cm^2) and 350 nJ for the second stage when pumped with 6.0 µJ (205 µm diameter, peak intensity of 190 GW/cm^2). The lower efficiency of the second stage is believed to be due to the group velocity mismatch between the UV pump and the visible seed. Both stages operating together typically yield a higher output energy (1.8 µJ) than the sum of the single stages. This is due to the fact that part of the amplified spectrum in the first stage overlaps with the amplification bandwidth of the second stage and therefore acts as a stronger seed than the supercontinuum alone, eventually leading to saturation in the second stage, as is also visible in the signal spectra shown in Fig. 4(a).

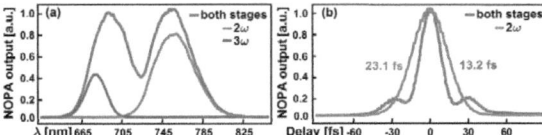

Fig. 4. (a) The spectral energy density of the signal pulse amplified by the SH and the TH of the pump laser (630 – 830 nm, red curve) is composed of the spectral region amplified only by the SH (690 – 830 nm, green curve) and the TH alone (630 – 715 nm, blue curve). These spectra allow for Fourier-limits of 11 fs, 19.7 fs and 28 fs, respectively. The measured spectral energy densities are normalized to the peak of the red curve. The corresponding measured autocorrelation traces with deconvoluted FWHM pulse durations are shown in (b).

After the two amplification stages the output signal is collimated with a spherical mirror and compressed using a sequence of fused silica prisms with an apex angle of 68.7°. Characterization of the pulses is performed online with a dispersion-free autocorrelator that provides direct information in the time domain [21] and alternatively with a ZAP-SPIDER setup that characterizes the pulses in the spectral domain [22]. We use a 30 µm thin BBO crystal for both devices. As auxiliary pulse for the ZAP-SPIDER we use the residual fundamental at 1025 nm after SHG and SFG, which is stretched by transmission through 1270 mm SF57 glass to a FWHM duration of 1.4 ps.

4. Results and discussion

4.1 NOPCPA on the mJ-level approaching the octave

The amplified signal spectrum in case of the mJ-level, cascaded two-color-pumped NOPCPA, which is outlined in section 2, ranges from 575 to 1050 nm with a total confined energy of 1.9 mJ. The pulse has a Fourier-limit of 4.5 fs (FWHM duration, see Fig. 2 inset) and a central wavelength of 782 nm. The spectral boundaries are determined by the effective phase-matching bandwidth shown in Fig. 5(a). The effective phase-mismatch in a BBO crystal of length L is calculated according to Ref. [15] as the sum of crystal-dependent wavevector-mismatch ΔkL and amplification-dependent OPA-phase (Eq. (8) in Ref. [15]). ΔkL leads to a phase-slippage between the pump wave, the seed wave and the idler wave generated in the BBO. The OPA-phase is a phase imprinted on the signal during amplification so as to compensate for the phase-slippage and therefore maintain high gain even in areas of significant ΔkL [15]. A maximum effective phase-mismatch of $\pm\pi$ is acceptable for coherent build-up of the amplified signal in the small-signal gain regime. The compensating effect of the OPA-phase broadens the acceptance bandwidth. In our experiment, the amplified bandwidth in the NOPCPA stage pumped by the SH matches exactly the calculated one shown in Fig. 5(a). In case of the TH-pumped stage, the measured amplified bandwidth is even slightly broader. This may be because Eq. (8) in Ref. [15] is only valid for low pump depletion.

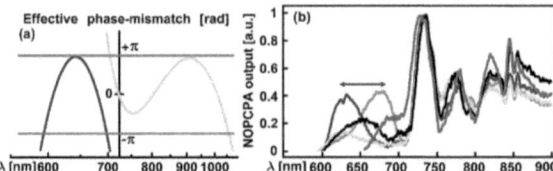

Fig. 5. (a) Effective phase-mismatch as function of seed wavelength including wave-vector-mismatch ΔkL and OPA-phase for the individual NOPCPA stages: $\lambda p = 532$ nm, $\theta = 23.62°$, $\alpha = 2.23°$ (green) and $\lambda p = 354.7$ nm, $\theta = 34.58°$, $\alpha = 3.40°$ (blue). The red horizontal lines label $\pm \pi$. (b) The spectral region below 700 nm is shapeable via adjusting the pump delay and the phase-matching angle (varied colors) in the NOPCPA stage pumped by 3ω.

Figure 5(b) shows a cut-out of signal spectra obtained through amplification by the SH and the TH. A steep spectral edge at around 700 nm leads to satellite pulses in the time domain after almost Fourier-limited compression [12], potentially degrading the temporal pulse contrast on the femtosecond timescale. This can be avoided by spectral shaping of the region below 700 nm via adjusting the pump delay and phase-matching angle in the NOPCPA stage pumped by the TH. Consequently, if one achieves adaptive compression of this octave-spanning pulse close to the Fourier-limit, optimization of the temporal structure of the compressed pulse seems to be possible via spectral shaping in the NOPCPA stages. The spectral structure above 700 nm in the signal spectrum in Fig. 2 is mainly dominated by modulation of the unamplified seed, which is spectrally broadened via self-phase modulation in the tapered hollow-core capillary fiber filled with neon gas.

It has been predicted that the overall pump-to-signal conversion efficiency and the full signal bandwidth of a NOPCPA are functions of group delay of the seed spectral boundaries [23]. Measured results for our cascaded two-color-pumped NOPCPA are summarized in Fig. 6 and represent to our knowledge the first reported measurement. Figure 6 shows that a variation from the seed GD of 37 ps, which is chosen for most of the measurements, can improve the conversion efficiency even further. A GD of (69 ± 2) ps for 575 to 1020 nm leads to the highest overall conversion efficiency of $(12.5 \pm 1.3)\%$, while still maintaining the signal bandwidth. The GD for 700 to 1020 nm and for 575 to 740 nm is roughly 33 ps and 43 ps, respectively. In this case, the conversion efficiency is 14.4% and 8.8% in the NOPCPA stages pumped selectively by the SH and the TH, which is the highest conversion efficiency obtained for the SH-pumped stage. In the SH-pumped stage, the pump energy is 16 mJ and the signal energy is 2.3 mJ. For the TH-pumped stage, the corresponding energies are 8 mJ and 0.7 mJ, yielding a total output energy of 3 mJ. Hence, both stages are operated near saturation. At the experimentally verified optimum seed GD, the results show a signal spectrum in the SH-pumped stage similar to that in Ref. [12] but with higher conversion efficiency, although the seed GD for 700 to 1020 nm is similar in both cases. We suspect that this is because we do not use an acousto-optic modulator (Dazzler) in the unamplified seed beam and for this reason observe a spatially and spectrally more homogeneous unamplified seed beam profile in the present experiment.

In general, the temporal overlap between the seed pulse and the 78 ps (FWHM) pump pulse increases with increasing seed GD until the optimum GD is reached. In this range, the conversion efficiency grows and the signal bandwidth stays constant as long as the NOPCPA stage is operated near saturation for most of the seed wavelengths. Otherwise, a decrease in seed intensity because of an enhanced stretching ratio can lead to lower signal bandwidth due to a lack of saturation [23]. Beyond the optimum GD, the seed pulse increasingly experiences the Gaussian temporal shape of the pump pulse, leading to a decrease in signal bandwidth and conversion efficiency.

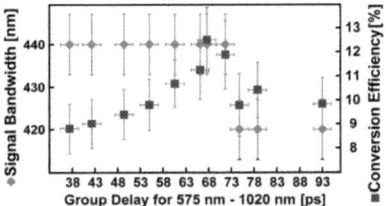

Fig. 6. Measured full signal bandwidth (red diamonds) and the overall pump-to-signal conversion efficiency (blue squares) as function of group delay between the seed spectral boundaries. The optimum group delay is found to be (69 ± 2) ps.

In conclusion, with a proper and precise dispersion management, two-color-pumped NOPCPA with type-I phase-matching in BBO is a promising approach for the generation of sub-two-cycle light pulses on the mJ-level. In this case, the present technique would close the gap between chirped-pulse amplifiers with HCFs (~4 fs, ~1 kHz, ~1 mJ [24,25]) and few-cycle one-color-pumped NOPCPA systems (~8 fs, ~10 Hz, 15-130 mJ [11,12]).

4.2 Proof-of-principle compressibility with high-repetition rate NOPA on the µJ –level

The amplified signal spectrum in case of the µJ-level cascaded two-color-pumped NOPA described in section 3 ranges from 630 to 830 nm with a total confined energy of 1.8 µJ. The spectral overlap is chosen to be similar to that in the mJ-level NOPCPA setup in section 2, with a central wavelength of 740 nm. Figure 4 shows the amplified spectrum and the corresponding measured autocorrelation trace (both in red). With identical prism compressor settings, we measure a FWHM pulse duration of 23.1 fs (Fourier-limit: 19.7 fs) and 95.0 fs (Fourier-limit: 28.0 fs) for the signal pulse amplified only by the SH and only by the TH. The signal pulse amplified in both stages is compressed to 13.2 fs, compared to its Fourier-limit of 11.0 fs. The measurement matches the simulation of our prism compressor and the compressibility is comparable to other experiments employing related prism compressors [20,26].

The prism compressor setting for optimum compression of the signal pulse resulting from amplification in both NOPA stages is also the optimum setting for the signal pulse amplified by the SH alone, in contrast to the signal amplified by the TH alone. To compress the signal amplified in the TH-pumped stage, the second prism of the compressor is further inserted. Consequently, less negative GDD is required if the signal amplified in the TH-pumped stage is to be compressed. During the course of the experiments, a deformable mirror as prism compressor end-mirror was sometimes found to reduce the outer wings of the compressed pulse.

A more detailed understanding can be gained from the determination of the spectral phase of the amplified pulse. Figure 7(a) shows the result of a ZAP-SPIDER [22] measurement for a particular pulse with a total amplified range of 615 to 780 nm. Figure 7(b) shows the amplified and compressed pulses calculated from the measured spectral intensity and phase given in Fig. 7(a). The FWHM pulse durations determined in this way are close to the results obtained with autocorrelation measurements of the same pulses.

The compressibility is determined by three factors. First, the bandwidth, whose phase can be managed throughout the present dispersive bulk materials (YAG and BBO crystals, fused silica lenses) and the fused silica prism-compressor, is limited. In case of the optimum compressor setting for the investigated spectral region a nearly vanishing spectral phase over a wide range results. The GD as a function of the signal wavelength shows a maximum around 710 nm, which limits the bandwidth of the compressed pulse. Furthermore, the third-order dispersion (TOD) due to the dispersive components leads to satellite pulses after compression.

These limitations mean that the spectral components below 660 nm and above 800 nm occur in the wings of the compressed pulse for the case of Fig. 4. This is responsible for the observation of the long pulse durations seen when the signal is amplified only by the TH. Nevertheless, the TH-pumped stage amplifies spectral components not amplified in the SH-pumped stage and whose phase can still be managed in our case using the prism compressor. For this reason, a shorter pulse duration is achieved in case of amplification in both NOPA stages compared to using only the SH-pumped stage, for the same prism compressor settings.

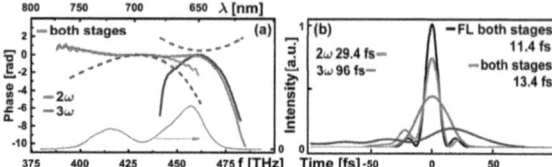

Fig. 7. (a) Spectral phase of the compressed signal pulses retrieved by the ZAP-SPIDER measurement. Solid red curve: amplification by the SH and TH of the pump laser, solid green: amplification only by the SH, solid blue: amplification only by the TH; all for identical settings of the compressor. For comparison the calculated OPA-phase for the two stages is shown with dashed lines. The solid grey curve shows the spectral energy density of the signal. Taking the measured spectral phase, spectral intensity and pulse energy leads to the retrieved pulses shown in (b) with their FWHM pulse duration.

Second, the phase imprinted on the signal to compensate for phase-mismatch and to maintain high gain (OPA-phase) comes into play [27]. As mentioned in subsection 4.1 and outlined in Ref. [15] and Ref. [28], this phase contribution affects ultra-broadband optical parametric amplification. Employing Eq. (8) of Ref. [15] for the case of small depletion, these phase contributions due to amplification in the TH-pumped and the SH-pumped NOPA stages are shown as dashed lines in Fig. 7(a) for the experimentally determined parameters like crystal angles and the pump intensities. One can see that the phase contributions have opposite signs. A positive linear chirp defined in the time domain leads to a parabola with a negative second derivative in the frequency domain via Fourier-transformation. In contrast to the SH-pumped NOPA stage, the phase imprinted on the signal during parametric amplification by the TH shows negative group delay dispersion (GDD). This matches our observation that less negative GDD was required with the prism compressor when the signal pulse was amplified by the TH alone. Moreover, these two phase contributions partially compensate each other and therefore lead to an extended region of reduced residual phase around the spectral overlap.

To experimentally verify this phase effect, we adjusted both NOPA stages to similar amplified spectra centered at 670 nm (different from the pulses previously described) and subsequently compressed the signal pulse amplified by each stage separately with the same prism compressor spacing. These compressed pulses were characterized with the autocorrelator. It was found that for optimum compression of the signal pulse amplified in the TH-pumped NOPA stage, the second prism in the compressor was inserted by about 2 mm more compared to the signal amplified in the SH-pumped stage. This implies that the prism compressor applies less negative GDD to the signal pulse, which matches the signs of the OPA-phase contributions due to parametric amplification shown in Fig. 7(a).

Note that the residual phase of the compressed signal pulse amplified by the SH or the TH alone is different from the residual phases of the corresponding spectral regions in case of amplification in both NOPA stages. Apart from the compensating effect of the phase-contribution due to parametric amplification in the region of spectral overlap, this occurs because Eq. (8) in Ref. [15] is strictly speaking only valid for the case of low pump depletion (i.e. negligible saturation). This observation has the consequence that the two amplification stages and the pulse compression can only be optimized simultaneously.

Third, cross-phase modulation (XPM), which is possible in optical parametric amplification as a result of the high pump intensities, would lead to positive GDD [29]. According to Ref. [29], we calculate the B-integral due to XPM in our NOPA stages to be 1.4 rad in the SH-pumped and 0.8 rad in the TH-pumped stages. This is comparable to the value found in the investigation of XPM [29]. Since the B-integral of the SH-pumped stage is approximately twice as high as that of the TH-pumped stage, a higher additional positive GDD can be present in the SH-pumped stage. This is consistent with our observations.

In conclusion, cascaded two-color-pumped NOPA with type-I phase-matching in BBO is also a promising approach for the generation of sub-two-cycle light pulses on the µJ-level in high-repetition rate NOPA systems.

5. Conclusion and perspectives

In this work results from two novel experimental setups were analyzed to investigate the feasibility of two-color pumping of a NOPCPA for the generation of high energy pulses approaching the single cycle regime. At low repetition rates, 3 mJ pulses with a nearly octave wide spectrum were demonstrated in a first double-stage arrangement. At 100 kHz repetition rate and µJ output energies the spectral phase and compressibility was studied. The two-color scheme exhibits a slowly and continuously varying spectral phase that should be well compensatable with existing compression schemes. The addition of the visible part of the pulse spectrum by the 3ω-pumping indeed shortens the pulse by nearly a factor of two without any change to the prism compressor. We conclude from the combination of results that the concept of two-color pumping can be expanded to multiple stages. Proper care has to be taken in the design and alignment to utilize the full available pump energy. Pulses approaching a J energy and a duration around 5 fs seem on the horizon. While sub-5s pulses were previously reported with 400 nm Ti:sapphire based pumping, this range is now reachable with the 1050 nm pump lasers of much higher energy.

Various challenges have to be resolved to achieve these ambitious goals. Already in the present experiments matching of the wavefront and beam pointing of the two contributions to the composite pulses needed high attention. We suspect that the birefringent nature of the amplifier crystals, possible inhomogeneities in the material, partial depletion of the pump and associated spatially-dependent OPA phase and classically neglected higher-order nonlinear interactions are the main causes of this situation. The OPA phase will depend selectively on the pump color, the degree of saturation and therefore on the OPA's location within an extended amplifier chain, and on the particular phase-matching adjustment of a selected crystal. Therefore an adaptive phase-correction will most likely be desirable for routine operation. An acousto-optic programmable dispersive filter with sufficient bandwidth is already available. Last but not least, the correct stretching ratio of seed and pump for the optimum balance of bandwidth, compressibility and overall efficiency is of high importance. For all these issues the present report provides a first basis.

Two-color pumping potentially allows more efficient usage of the available pump energy. Only for the short wavelength part of the output spectrum are the "expensive" short wavelength pump photons used, while the red part of the spectrum is amplified with the help of the remaining green pump light. Already now the generation of the UV pump utilizes the fundamental pump pulses twice. In future extensions of the concept one could even think of adding another amplifier stage pumped by the residual pump fundamental to widen the spectrum in the near infrared and to add even more energy.

It is interesting to compare the optical principles underlying the present approach for the generation of extremely broadband pulses with other methods of light wave synthesis. Recently it was demonstrated that interferometric spatial addition of phase-locked pulses generated at neighboring wavelength ranges in a fiber-based MHz system leads to single cycle near-infrared pulses [30]. This approach of individual amplification and dedicated compression of the spectral parts allows for more flexibility. On the other hand, the spatially and temporally stable overlap of the contributions, without adverse effects from inhomogeneities in scaling to high pulse energies and consequently large beam sizes, might be critical. In our approach a common seed and beam path is used and could eventually be more practical.

The motivation for our work was the high energy, low repetition rate regime. In the course of the work we realized that two- or even multiple color pumping for the extension of the output spectrum is also feasible at 100 kHz repetition rates. This might provide interesting sources for spectroscopic investigations of samples in the condensed phase that require only low pulse energies. It is certain that the concept can close existing gaps between Ti:sapphire based systems with hollow-core capillary-fiber compression (~4 fs, ~750 nm, ~1 mJ, ~1 kHz [24,25]) and 8 fs, 800 nm, 10 Hz, 15-130 mJ NOPCPA systems [11,12].

Acknowledgements

This work was supported by Deutsche Forschungsgemeinschaft (contract TR18), the association EURATOM-Max-Planck-Institut für Plasmaphysik, the Cluster of Excellence Munich-Centre for Advanced Photonics (MAP) and by the Austrian Science Fund within the Special Research Program F16 (Advanced Light Sources). We acknowledge support by the cooperation with the King-Saud-University. D. H. is grateful to the Studienstiftung des deutschen Volkes. The International Max Planck Research School on Advanced Photon Science (C. H.) is gratefully acknowledged. The authors thank N. Krebs for valuable support during the SPIDER measurements.

Appendix B6

Approaching the Full Octave: Noncollinear Optical Parametric Chirped Pulse Amplification with Two-Color Pumping

Daniel Herrmann, Christian Homann, Raphael Tautz, Laszlo Veisz, Ferenc Krausz, and Eberhard Riedle

Reprinted with permission from

Advanced Solid-State Photonics, OSA Technical Digest (CD) (Optical Society of America, 2011), paper JWC1.

http://www.opticsinfobase.org/abstract.cfm?URI=ASSP-2011-JWC1

Copyright © 2011 The Optical Society of America

Approaching the Full Octave: Noncollinear Optical Parametric Chirped Pulse Amplification with Two-Color Pumping

Daniel Herrmann[1,2], Christian Homann[1], Raphael Tautz[2,3], Laszlo Veisz[2], Ferenc Krausz[2,4], and Eberhard Riedle[1]

[1] *LS für BioMolekulare Optik, Ludwig-Maximilians-Universität München, Oettingenstr. 67, 80538 München, Germany*
[2] *Max-Planck-Institut für Quantenoptik, Hans-Kopfermann-Strasse 1, 85748 Garching, Germany*
[3] *present address: LS für Photonik und Optoelektronik, LMU München, Amalienstr. 54, 80799 München, Germany*
[4] *LS für Laserphysik, LMU München, Am Coulombwall 1, 85748 Garching, Germany*
E-mail:d.herrmann@physik.uni-muenchen.de

Abstract: We amplify ultrabroadband spectra to mJ energies: 575-1050nm by two-color-pumping and 675-1000nm by two-beam-pumping. We demonstrate the compressibility of these spectra and reveal the significance of a parametric phase imprinted on the signal.

1. Introduction and motivation

Few-cycle light pulses are of large interest for numerous applications in high-field science like the generation of surface high harmonic radiation and in time-resolved ultrafast spectroscopy of charge transfer in molecular systems. Noncollinear optical parametric chirped pulse amplification (NOPCPA) provides a powerful method to generate such pulses because of its broad amplification bandwidth, spectral tunability and high single-pass gain. Current state of the art multi-terawatt systems are mainly based on Nd^+ and Yb^+-based pump lasers and BBO as amplifying crystal and are capable of generating pulses down to 7.9 fs duration with energies of up to 130 mJ [1]. A further reduction of the pulse duration is not limited by the seed spectrum (500–1050 nm), but by the amplification bandwidth of BBO when pumping with the second harmonic (SH, ~ 532 nm) of the the pump laser. A possible solution for broader amplified spectra is to use other amplification crystals with broader gain spectra (e.g. DKDP), however they typically possess much lower conversion efficiencies than BBO.

We present the use of two different pump wavelengths with neighboring optimal spectral amplification regions for two subsequent stages of a NOPCPA chain. We also outline the usage of two pump beams with identical wavelength in one NOPCPA stage. Both concepts aim to achieve a much broader overall bandwidth.

In case of two-color-pumping, a first NOPCPA stage is pumped by the SH, and a second stage by the third harmonic (TH, 355 nm) of a Nd:YAG pump laser, we achieve an overall amplification from 575–1050 nm (Fig. 1), considerably broader than the 700–1000 nm spectral region reached before [1]. In this way we achieve signal spectra with Fourier-Limits of 4.5 fs. To test if potential problems like phase discontinuities arise in the spectral overlap region, we performed a proof-of-principle experiment with slightly narrower amplified signal spectra employing a low-power high repetition rate laser system of similar fundamental wavelength. We show that we can compress a composed spectrum to 13.2 fs by the use of a prism sequence only, close to its Fourier-Limit of 11 fs, and considerably shorter than the single signal spectra of either stage permit.

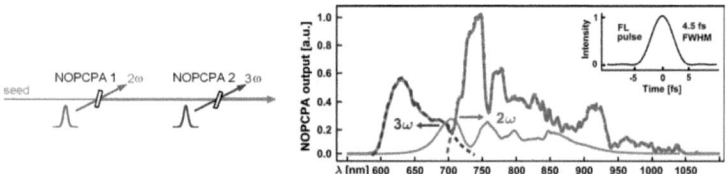

Fig. 1. By pumping the first stage of a cascaded NOPCPA chain by the SH of the Nd:YAG pump laser and the second stage by the TH, the spectral energy density of the signal pulse (575–1050 nm, red solid curve) allows for a Fourier-limit of 4.5 fs (inset). This spectrum is composed of the spectral region amplified only by the SH (700–1050 nm, green dotted curve) and the TH alone (575–740 nm, blue dotted curve). The seed spectrum (not to scale) is shown as gray solid curve.

For two-beam-pumping, a single NOPCPA stage is pumped by two beams at 532 nm under different phase-matching conditions resulting in two complementary gain bandwidths. The composed amplified signal spectrum ranges from 675–1000 nm and is compressed to 7.1 fs, which is close to the 6.7 fs Fourier-limit.

Both studies reveal and quantify the significant influence of a parametric phase imprinted on the signal and proof that a smooth spectral phase in the overlap region of the composed spectra is present employing these two schemes for optical parametric amplification. This opens up the route to a sub-5 fs multi-mJ NOPCPA system.

2.1 Cascaded two-color pumped NOPCPA as a novel technique for high-power two-cycle light pulses

To show the potential of two-color pumping for high-power few-cycle pulse generation, we extended and modified part of the Light-Wave-Synthesizer 20 (LWS-20) NOPCPA system [1,2]. The system consists of a Ti:Sa master oscillator from which the seed and pump pulses for the NOPCPA chain are derived. For the seed, part of the oscillator output is amplified in a 9-pass Ti:Sa multipass amplifier, delivering 25 fs, 800 µJ pulses at 1 kHz. These pulses are broadened in a tapered hollow-core fiber filled with neon at a pressure of 2 bar to a spectrum covering 500 – 1050 nm. The pulses are variably stretched in glass to roughly match the associated pump pulse durations and to study the NOPCPA efficiency as function of seed group delay (GD). We also attenuate the seed to a pulse energy of 4 µJ with neutral density filters, to simulate the expected pulse energy when a more complex stretcher consisting of a combination of grisms and an acousto-optic modulator is used, which will be necessary for a dispersion management allowing full compression of the generated spectra.

The remaining 40% of the oscillator output is used to seed the Nd:YAG pump laser, providing all-optical synchronization. This pump laser delivers two beams of 78-ps (FWHM), up to 1-J pulses at 1064 nm and 10 Hz repetition rate. Pulses at 532 nm are generated via type-II second-harmonic generation (SHG) of two fundamental 1064 nm beams in a 10 mm long DKDP crystal. The remaining fundamental and a part of the SH are relay-imaged onto a 10 mm long DKDP crystal for type-II sum-frequency generation (SFG). Typically 6 mJ at 532 nm and 14 mJ at 1064 nm are used to generate 8 mJ pulses of the TH at 355 nm. The beam diameters are 5 mm each. A high energy conversion efficiency of 40% and a TH quantum efficiency of 90% relative to the SH is observed. The SH and TH pump pulse durations are also 78 ps due to strong saturation in the conversion.

For the first NOPCPA stage 14 mJ of the 532 nm beam are split off and relay imaged onto a 5 mm long type-I BBO crystal, cut at 24°. With a seed group delay of 37ps (575–1020nm) and an internal noncollinear angle of 2.2° we achieve broadband amplification from 700-1050 nm (green curve in Fig. 1) with an energy of 1.4 mJ. The output signal of this stage is then further amplified in a second NOPCPA stage pumped by the third harmonic in a 3 mm long BBO crystal, cut at 32.5°. With a noncollinearity angle of 3.4°, we amplify a spectral region from 575–740 nm (blue curve in Fig.1), which neatly connects to the amplified spectrum of the first stage. The resulting pulses have a total energy of 1.9 mJ, a spectral range of 575–1050 nm (Fig. 1), and a Fourier-Limit of 4.5 fs [2]. Enhanced stretching of the seed to a GD of (69 ± 2) ps for 575 to 1020 nm leads to 3 mJ total signal energy with the highest overall pump-to-signal conversion efficiency of $(12.5 \pm 1.3)\%$, while maintaining the signal bandwidth. The GD for 700–1020 nm and for 575–740 nm is then 33 ps and 43 ps.

2.2 Compressibility of composed spectra

Compression of the single spectra of the SH pumped stage is already shown [1], and accordingly no difficulties are expected for the TH pumped stage alone. Crucial is the spectral overlap region and the question if phase discontinuities arise, e.g., by different parametric phases in the stages due to the different pump wavelengths [3].

To test this, we performed a similar experiment with a Yb-based pump laser system, delivering 300 fs, 40 µJ pulses at a center wavelength of 1025 nm and a repetition rate of 100 kHz. We accordingly built two cascaded BBO based noncollinear OPA stages pumped by the SH and TH of the pump laser [4]. As seed light we used a supercontinuum generated in a 2 mm thick YAG plate. We amplified spectral regions from 690–830 nm with the SH and from 630–715 nm with the TH (Fig. 2). Special care had to be taken to spatially overlap the output of the two stages, which was not ensured automatically. The pulse was then compressed in a fused silica prism compressor and the output characterized by a dispersion-free autocorrelator and a ZAP-SPIDER setup.

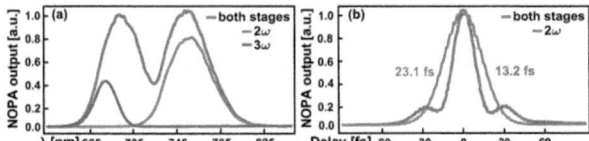

Fig. 2. Amplified signal spectra (a) and autocorrelation traces (b) of the 2ω pumped NOPA stage (green), the 3ω pumped stage (blue) and of both stages operating simultaneously (red). These spectra allow for Fourier-limits of 19.7 fs, 28 fs, and 11 fs. The measured spectra are normalized to the peak of the red curve.

With the identical compressor setting we measure a pulse duration of 23.1 fs for the signal pulse amplified by the SH pumped stage alone and a duration of 95 fs for the 3ω pumped stage. Both stages together render a pulse duration of 13.2 fs, close to the Fourier-Limit of the complete spectrum of 11 fs [2]. Moreover, amplification of the same signal spectrum centered at 670 nm in each stage separately and subsequent compression reveal, that a parametric phase of strongly pump-wavelength-dependent magnitude and even sign as well as spectral and spatial behavior is imprinted on the signal during amplification.

3. Investigation of two-beam-pumped NOPCPA

To investigate the potential of two-beam pumping (TBP) for high-power few-cycle pulse generation, we extended the LWS-20 system [1,5]. As seed source we use 2 µJ pulses with a spectrum ranging from 650–1000 nm and a GD of 42 ps. The seed is negatively chirped by a reflection grism-pair and an acoustooptic modulator (Dazzler). The pump pulses are generated with the above described Nd:YAG laser amplifier. The TBP-NOPCPA stage consists of a 5 mm long type-I BBO crystal and is pumped with a variable total pump energy up to 12 mJ in two beams with different noncollinear angles (α_1=2.23°, α_2=-2.16°) included with the seed. The pump beam is relay imaged from the pump laser output on the BBO crystal and split into two parts with a 50%:50% beamsplitter. Each pump has its own delay stage for optimal temporal overlap with the seed.

The composed amplified signal pulse spectrum ranges from 675–1000 nm leading to a Fourier limit of 6.7 fs with a total energy of 0.5 mJ. The two pump pulses amplify spectral components of 700–1000 nm and 675–710 nm plus 800–1000 nm, respectively. We measure the signal energy as function of pump intensity and calculate the total pump-to-signal conversion efficiency. It turns out, that the conversion efficiency is comparable for TBP and one-beam-pumping with the same total pump energy. The amplified signal beam is compressed in a hybrid bulk/chirped mirror compressor in a vacuum chamber. A compressed pulse duration of 7.1 fs (Fig. 3(b)) is measured with a home-built all-reflective single-shot second-order autocorrelator [5]. It is also verified, that a pump interference pattern present in the BBO crystal plane is not transferred to the signal beam, no degradation of the spatial quality is observed.

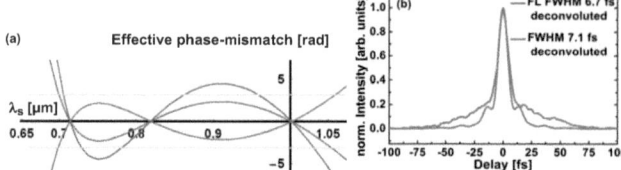

Fig. 3. (a) Effective phase-mismatch (purple), wave-vector mismatch ΔkL (blue) and parametric phase (red) as function of seed wavelength for a one-beam-pumped NOPCPA stage: λp = 532 nm, θ = 23.62°, α = 2.22°, L=5 mm. Green horizontal lines: ±π. (b) Autocorrelation for the measured (red) and Fourier-limited pulse(gray).

The effective phase-mismatch in a BBO crystal of length L is calculated according to Ref. [4] as the sum of crystal-dependent wave-vector mismatch ΔkL and amplification-dependent parametric phase ("OPA-phase"). ΔkL leads to a phase-slippage between the pump, the seed and the idler wave generated in the BBO. The present experiments reveal that the OPA-phase is indeed imprinted on the signal during amplification. The theoretical studies in Fig. 3(a) and Ref. [5,6] also show, that this parametric phase partially compensates the phase-slippage and therefore maintains high gain even in areas of significant ΔkL. A maximum effective phase-mismatch of $\pm\pi$ is acceptable for coherent build-up of the amplified signal in the small-signal gain regime. The compensating effect of the OPA-phase broadens the acceptance bandwidth. In our experiment, the amplified bandwidth in the NOPCPA stage pumped by the SH matches exactly the calculated one shown in Fig. 3(a) (purple).

4. Conclusion and outlook

Both demonstrated novel NOPCPA schemes lead to the amplification of extremely broadband composed spectra on the mJ-level with the potential of upscaling the energy. It is shown, that composed spectra from an OPA can be compressed close to their Fourier-limit even in the range of only a few optical cycles. This demonstrates that no phase discontinuities arise in the spectral overlap region. Pulses approaching a J energy and a duration around 5 fs seem on the horizon. Further realizations of our pumping concept can be envisioned, as for example combined ω and 2ω pumping for broadband amplification in the near-infrared spectral region.

5. References

[1] D. Herrmann, L. Veisz, R. Tautz, F. Tavella, K. Schmid, V. Pervak, and F. Krausz, "Generation of sub-three-cycle, 16 TW light pulses by using noncollinear optical parametric chirped-pulse amplification," Opt. Lett. 34(16), 2459–2461 (2009).
[2] D. Herrmann, C. Homann, R. Tautz, M. Scharrer, P. St.J. Russell, F. Krausz, L. Veisz and E. Riedle, "Approaching the full octave: Noncollinear optical parametric chirped pulse amplification with two-color pumping," Optics Express, 18(18), 18752–18762 (2010).
[3] I. N. Ross, P. Matousek, G. H. C. New, and K. Osvay, "Analysis and optimization of optical parametric chirped pulse amplification," J. Opt. Soc. Am. B 19(12), 2945–2956 (2002).
[4] C. Homann, C. Schriever, P. Baum, and E. Riedle, "Octave wide tunable UV-pumped NOPA: pulses down to 20 fs at 0.5 MHz repetition rate," Opt. Express 16(8), 5746–5756 (2008).
[5] D. Herrmann, R. Tautz, F. Tavella, F. Krausz, and L. Veisz, "Investigation of two-beam-pumped noncollinear optical parametric chirped-pulse amplification for the generation of few-cycle light pulses," Opt. Express 18(5), 4170–4183 (2010).
[6] A. Renault, D. Z. Kandula, S. Witte, A. L. Wolf, R. Th. Zinkstok, W. Hogervorst, and K. S. E. Eikema, "Phase stability of terawatt-class ultrabroadband parametric amplification," Opt. Lett. 32(16), 2363–2365 (2007).

Appendix B7

Role of Structural Order and Excess Energy on Ultrafast Free Charge Generation in Hybrid Polythiophene/Si Photovoltaics Probed in Real Time by Near-Infrared Broadband Transient Absorption

Daniel Herrmann, Sabrina Niesar, Christina Scharsich, Anna Köhler, Martin Stutzmann, and Eberhard Riedle

Reprinted with permission from
Journal of the American Chemical Society 133, 18220 (2011).

DOI: 10.1021/ja207887q

http://pubs.acs.org/doi/abs/10.1021/ja207887q

Copyright © 2011, American Chemical Society.

Role of Structural Order and Excess Energy on Ultrafast Free Charge Generation in Hybrid Polythiophene/Si Photovoltaics Probed in Real Time by Near-Infrared Broadband Transient Absorption

Daniel Herrmann,[†] Sabrina Niesar,[‡] Christina Scharsich,[§] Anna Köhler,[§] Martin Stutzmann,[‡] and Eberhard Riedle[*,†]

[†]Lehrstuhl für BioMolekulare Optik, Ludwig-Maximilians-Universität München, Oettingenstraße 67, 80538 München, Germany
[‡]Walter Schottky Institut, Technische Universität München, Am Coulombwall 4, 85748 Garching, Germany
[§]Lehrstuhl EP II, Universität Bayreuth, Universitätsstraße 30, 95440 Bayreuth, Germany

ⓈSupporting Information

ABSTRACT: Despite the central role of light absorption and the subsequent generation of free charge carriers in organic and hybrid organic–inorganic photovoltaics, the precise process of this initial photoconversion is still debated. We employ a novel broadband (UV–Vis–NIR) transient absorption spectroscopy setup to probe charge generation and recombination in the thin films of the recently suggested hybrid material combination poly(3-hexylthiophene)/silicon (P3HT/Si) with 40 fs time resolution. Our approach allows for monitoring the 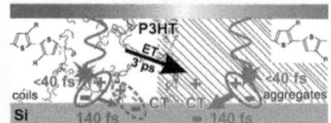 time evolution of the relevant transient species under various excitation intensities and excitation wavelengths. Both in regioregular (RR) and regiorandom (RRa) P3HT, we observe an instant (<40 fs) creation of singlet excitons, which subsequently dissociate to form polarons in 140 fs. The quantum yield of polaron formation through dissociation of delocalized excitons is significantly enhanced by adding Si as an electron acceptor, revealing ultrafast electron transfer from P3HT to Si. P3HT/Si films with aggregated RR-P3HT are found to provide free charge carriers in planar as well as in bulk heterojunctions, and losses are due to nongeminate recombination. In contrast for RRa-P3HT/Si, geminate recombination of bound carriers is observed as the dominant loss mechanism. Site-selective excitation by variation of pump wavelength uncovers an energy transfer from P3HT coils to aggregates with a $1/e$ transfer time of 3 ps and reveals a factor of 2 more efficient polaron formation using aggregated RR-P3HT compared to disordered RRa-P3HT. Therefore, we find that polymer structural order rather than excess energy is the key criterion for free charge generation in hybrid P3HT/Si solar cells.

1. INTRODUCTION

Thin film solar cells constitute one of the future technological solutions for sustainable energy supply. A particularly promising route is offered by solar cells made from organic semiconductors or inorganic semiconducting nanoparticles.[1−4] In recent years, hybrid solar cells based on an organic semiconductor in conjunction with an inorganic nanoscale material are considered as an alternative to purely organic solar cells, as they allow achieving additional functionality by combining the advantages of the two materials.[5−9] In comparison to polymers, inorganic semiconductors offer a broader spectral range of absorption, particularly in the NIR spectral range, a higher charge carrier mobility, and a better thermal and morphological stability. At the same time, their application in the form of nanoparticles enables the possibility of band gap tuning for sufficiently small nanoparticle diameters and the technological advantages of purely organic solar cells, such as low-cost solution processing, roll-to-roll assembly, or processing onto flexible substrates, are maintained. In the literature, various composites based on Si, ZnO, TiO_2, CdSe, and a few other nanomaterials are currently of scientific interest.[6−15]

However, in hybrid and organic solar cells, a detailed fundamental understanding of the processes of light absorption, formation of free polarons, and the subsequent transport of these charges to the electrodes, which are central to their operation, is lacking. In purely organic solar cell devices, these photophysical processes have been shown to depend strongly on the morphology of the heterojunction[16] so that the device efficiency can be improved significantly by the processing conditions of the film.[17−19] The widely used and highly attractive polymer poly(3-hexylthiophene) (P3HT) can form two distinct morphological phases associated with different chain conformations. If the P3HT chain adopts a random coil conformation, the resulting film is amorphous. The associated absorption spectrum is unstructured with a maximum centered around 450 nm (about 2.8 eV). This disordered structure prevails for regiorandom P3HT (RRa-P3HT). In regioregular P3HT (RR-P3HT), the polymer chains can planarize and assemble to form weakly coupled H-aggregates,[20,21] which arrange in closely (a few angstroms) packed two-dimensional lamellar structures via π-stacking.[22−25] Their spectroscopic signature is a well-structured absorption spectrum with a 0–0 vibronic peak around 600 nm (about 2.0 eV).[21]

Received: June 1, 2011
Published: September 26, 2011

Such aggregates are partially formed when RR-P3HT is embedded in a poor solvent or in a film after spin-coating from solution. RR-P3HT is a semicrystalline polymer whose degree of crystallization can be controlled by processing conditions. While it became clear that the charge carrier mobility is enhanced in aggregated P3HT chains,[26,27] studies on the role of the aggregated or coiled conformation in the process of charge carrier generation and separation in organic devices have been emerging only recently.[16]

For the design and operation of a solar cell, it is therefore of crucial importance to understand the influence of morphology on each of the individual photophysical steps. In organic semiconductors, there is widespread agreement about the photoconversion process. The elementary step is light absorption to generate excited states of a donor, followed by diffusion of the excitation to the internal interface formed by a donor adjacent to an acceptor and the decisive electron transfer from the excited donor to the acceptor forming a Coulombically bound electron−hole pair. Ideally, this is followed by their dissociation into free charges that move away from the interface, preferentially not suffering bimolecular recombination before being collected at the respective electrodes. Moreover, there is agreement that in purely organic blends the process of charge carrier generation takes place on an ultrafast time scale in the range of 100 fs.[16,28−31]

However, the exact mechanism of charge separation is still debated for purely organic solar cells and still in an early phase for hybrid composites. For P3HT in combination with [6,6]phenyl-C_{61}-butyric acid methyl ester (PCBM), which has been demonstrated to achieve power conversion efficiencies of around 5%,[32] there are suggestions that Frenkel-type excitons are the primary photoexcitations that dissociate into free charges.[16,29,30] There are indications that the charge separation takes place more efficiently for blends of RR-P3HT:PCBM than for RRa-P3HT:PCBM.[16] In contrast, prompt polaron formation during laser excitation was considered for neat P3HT and RRa-P3HT:PCBM.[30,33] For the polymer PCDTBT in combination with PCBM, there are also contradicting interpretations in discussion. On the one hand, it has been suggested that light absorption may directly create mobile electrons and holes by interband $\pi-\pi^*$ transitions which would subsequently evolve into Coulombically bound excitons in less than 1 ps.[31] Similarly, for a composite of a PPV derivative with PCBM, the primary photoexcitation has been suggested to be an ultrafast electron transfer on the time scale of 45 fs.[28] On the other hand, ultrafast exciton dissociation to form free charges was also considered very recently for PCDTBT:PCBM blends.[34] For hybrid donor−acceptor materials, detailed ultrafast spectroscopic investigations are still in an early phase. The question, whether light absorption initially creates free charge carriers or excitons, is central to the understanding of light harvesting in organic and hybrid systems. A problem in resolving this issue pertains to the experimental limits of time resolution and spectral range that are accessible to optical probing. Here, we have developed a novel ultrabroadband transient absorption spectroscopy setup with a time resolution of 40 fs covering the entire broad spectral range from 415 to 1150 nm without interruption. This allows us to monitor both the kinetics of the decay of the primary excitation and its evolution into a charge pair state in thin hybrid films.

In our studies, we focus on composites of P3HT in combination with silicon which is a particularly promising inorganic acceptor for several reasons. It unifies an almost unlimited abundance with environmentally friendliness, allowing for its widespread use. Silicon additionally provides high electron affinity and allows for rapid electron delocalization and screening after charge transfer which may prevent back transfer and enables fast transport away from the interface. Because of the higher dielectric constant of silicon compared to PCBM, this effect should be even more pronounced as in purely organic films, thus rendering silicon a very promising alternative to PCBM for photovoltaic devices and fundamental studies. Proof-of-principle investigations of charge transfer in hybrid P3HT/Si systems were recently performed using electron spin resonance (ESR).[9] Prototype devices exhibited a relatively high open-circuit voltage of 0.75 V,[10] and power conversion efficiencies of around 1% have been achieved.[8] However, no detailed spectroscopic understanding has been available so far. Because of its current availability in crystalline, nanocrystalline, and amorphous forms, silicon serves as a model system with fundamental implications for various other hybrid or organic material systems. In particular, the exciton dissociation mechanism and therefore the photophysics of charge generation and separation can be studied more clearly in silicon-based devices. The reason for this is that these processes are difficult to assess in the widely studied P3HT/PCBM composite, since the various PCBM transient signals from the visible to NIR[35] superimpose with the transient absorption by the polaron and exciton of P3HT.

In our study, we employ Si nanocrystals (Si-ncs) and polycrystalline silicon (poly-Si) as the electron acceptor in order to study both film geometries of interest, bulk heterojunctions and planar heterojunctions. For the development of efficient commercial solar cells, the bulk heterojunction structure is favored, since it offers a particularly large donor−acceptor interface. The efficiency of planar heterojunctions is limited by the smaller interfacial area, yet the two-dimensional interface area avoids recombination associated with interrupted percolation pathways and cross-currents of electrons and holes. However, as our studies primarily focus on the fundamental principles of the photophysics (charge generation and separation), our devices are optimized for an unambiguous data interpretation. To address the dependence of charge carrier separation on polymer structural order, both disordered RRa-P3HT and semicrystalline RR-P3HT were used. Our pump−probe setup allows us to directly monitor in real time the process of charge generation in hybrid thin film P3HT/Si heterojunctions, here at room temperature in the absence of an applied external field. For both RR- and RRa-P3HT, we observe an instant creation of singlet excitons that subsequently dissociate to form polarons on an ultrafast time scale. We observe that the yield of polaron formation through exciton dissociation is significantly enhanced by adding Si as electron acceptor. Furthermore, we find that the yield of polaron formation and the degree of Coulombic binding of the corresponding polaron pairs formed in P3HT/Si depend on the polymer structural order, with efficient free charge carrier generation in RR-P3HT/Si and geminately bound charge carriers formed in RRa-P3HT/Si.

2. SAMPLE PREPARATION AND SUMMARY OF EXPERIMENTAL METHODS

2.1. Sample Preparation. In this work, three different types of poly(3-hexylthiophene) with varying degree of aggregation were used. For fundamental studies of interactions between disordered and ordered regions of P3HT, films of a RR-P3HT (BASF SE, Sepiolid P 100, regioregularity ∼95%, M_w = 50 000 g/mol, M_w/M_n = 2.2) were prepared by spin-coating from chloroform ($CHCl_3$) solutions with a concentration of 7.5 mg/mL onto precleaned glass substrates. The same polymer was dissolved in spectroscopically pure chloroform, toluene, and 1,2-dichlorobenzene with concentrations of 0.075 mg/mL and filled

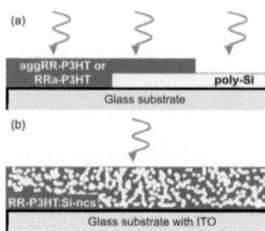

Figure 1. Sample architecture of hybrid P3HT/Si thin film: (a) planar heterojunction and (b) bulk heterojunction on glass substrates.

Table 1. List of Samples Used in This Work along with Their Abbreviations

sample	abbreviation
RR-P3HT (BASF), spun from CHCl$_3$	RR-P3HT
RRa-P3HT (Bayreuth), spun from CHCl$_3$	RRa-P3HT
RR-P3HT (Rieke), spun from CHCl$_3$/EtAc	aggRR-P3HT
silicon nanocrystals	Si-ncs
polycrystalline silicon	poly-Si
RR-P3HT:Si-ncs bulk heterojunction	RR-P3HT:Si-ncs BHJ
RRa-P3HT/poly-Si planar heterojunction	RRa-P3HT/poly-Si PHJ
aggRR-P3HT/poly-Si planar heterojunction	aggRR-P3HT/poly-Si PHJ

in 1 mm fused silica cuvettes for the transient absorption spectroscopy of P3HT in solution.

P3HT/Si thin film heterojunctions with varied morphology were prepared as planar heterojunctions (PHJs, Figure 1a) and bulk heterojunctions (BHJs, Figure 1b). For the PHJs with varied polymer structural order, RRa-P3HT (University of Bayreuth, Germany, M_w = 40 000 g/mol, M_w/M_n = 2.4) dissolved in chloroform (5 mg/mL), and RR-P3HT (Rieke Metals, M_w = 39 000 g/mol, M_w/M_n = 2.0) dissolved in 98% chloroform and 2% ethyl acetate (EtAc) (5 mg/mL) was used. The former was synthesized by treating thiophene with FeCl$_3$.[36] For the latter, ethyl acetate as nonideal solvent for P3HT was admixed to enhance the aggregation of the RR-P3HT molecules,[37−39] which is therefore referred to as aggRR-P3HT. All sample preparations were performed under argon atmosphere.

Silicon nanocrystals (Si-ncs) were synthesized in a low-pressure microwave plasma reactor by decomposition of silane.[40] The pressure of the process gases and the microwave power determine the mean diameter of the Si-ncs. Phosphorus doping of the Si-ncs was achieved by adding phosphine during growth and increases the carrier mobility in the Si-ncs. The nominal doping concentration is defined by the phosphine flow and the flow of the total precursor gas. The Si-ncs used in this work have a mean diameter of 4 and 18 nm and nominal doping concentrations of 5×10^{20} and 6.5×10^{19} cm^{-3}, respectively. The standard deviation of the particle diameter is typically $\sigma \approx 1.4$ nm.

Blend films of RR-P3HT and Si-ncs with a weight ratio of 5:1 were spin-coated under nitrogen atmosphere from solutions in chloroform (concentration 7.5 mg/mL) to form RR-P3HT:Si-ncs BHJs (Figure 1b). For the intended application as a solar cell, 15 mm × 15 mm × 1.1 mm aluminoborosilicate with an approximately 110 nm thick conductive indium tin oxide layer (ITO, Delta Technologies, R_s = 5−15 Ω) was used as a substrate. Before spin-coating, the substrates were cleaned by subsequent ultrasonic treatment in acetone and isopropanol for 10 min each. A typical sample layer thickness of 100 nm was achieved. The thin film samples were sealed against air by using fused silica coverslips (150 μm thin) and silicone sealant.

The refractive index of the Si-ncs was previously measured to be about 2.0 with only a slight monotonic decrease with wavelength. The refractive index of the blend films can be expected to be close to the one of a neat P3HT film, which has a reported index of around 1.7−2.0.[41] Consequently no significant change of the Fresnel losses upon mixing of the two materials is expected and the weak excitation should also not lead to a transient change.

Polycrystalline silicon (poly-Si) films were prepared by silver-induced layer exchange (AgILE).[42] For a resulting poly-Si film thickness of 30 nm, an amorphous silicon precursor layer (50 nm) was grown on top of a 30 nm silver layer on a fused silica substrate. The crystallization was performed at 800 °C for 10 h under nitrogen atmosphere. Afterwards, the silver was etched away with a 1:1 mixture of hydrogen peroxide and ammonia solution at 100 °C. We prepared PHJs of 40 nm RRa-P3HT or aggRR-P3HT and 30 nm poly-Si (aggRR-P3HT/poly-Si PHJ and RRa-P3HT/poly-Si PHJ, Figure 1b) under nitrogen atmosphere. To be able to perform TA spectroscopy of the individual materials and of the heterojunctions, the poly-Si covers only a part of the substrate (Figure 1a). Furthermore, we removed a stripe of the P3HT on the Si side using a cotton tip with chloroform. A summary of the sample structures used is given in Table 1.

2.2. Summary of Experimental Methods. Details about the basic optical and morphologic characterization of the thin film samples and the novel ultrafast transient absorption spectrometer are given in the Supporting Information. To investigate the nature of the photoexcitations and their inherent kinetics after visible excitation, we used an ultrafast 1 kHz pump−probe setup with a probe range of 290−740 nm.[43] The visible excitation with 15 fs pulses in the range from 450 to 720 nm is accomplished by a noncollinear optical parametric amplifier (NOPA).[44,45] We expand the probe range to 415−1150 nm by the combination of two supercontinuum generation stages and a NIR-OPA operated at 1180 nm, allowing for broadband UV−Vis-NIR TA spectroscopy with 40 fs time resolution (Figure S3). This OPA was seeded with a supercontinuum from a YAG crystal[46] and generated the Vis−NIR probe continuum in a CaF$_2$ crystal. The pump and probe were focused towards the sample to a 210 and 110 μm $1/e^2$ beam diameter that allows ensemble averaging over the finely grained morphology of the thin films to mimic their usage as photovoltaic device and to ensure low local excitation densities.

3. RESULTS AND DISCUSSION

3.1. Structural Implications of Regioregularity and Solvent. In our study, we aim to understand the process of charge carrier generation in the polymer P3HT and in the hybrid system formed by P3HT in combination with silicon. P3HT is a semicrystalline polymer whose degree of aggregation depends on the degree of regioregularity of the chain as well as on the solvent used. In order to assess the role of aggregated P3HT chains in the charge generation process, we employed samples with different degrees of regioregularity and spun from different solvents, as summarized in Table 1. The resulting structure of the films was carefully monitored via atomic force microscopy (AFM). The corresponding topographical images in 2D and 3D plots are shown in Figures 2a−c and Figure S1 and reveal a varied aggregation: While RRa-P3HT exhibits a rather long spatial coherence length (Figure 2a, Figure S1a), aggRR-P3HT possesses a fine structure (Figure 2c, Figure S1b). RR-P3HT is between these two extremes (Figure 2b).

The differences can be understood by regarding the different processing conditions. In the case of aggRR-P3HT, we have used chloroform as the main solvent, which exhibits a lower (61 °C)

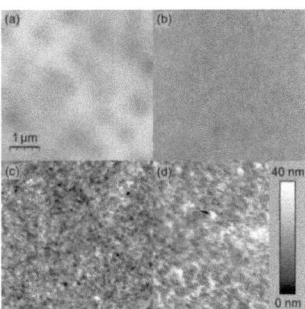

Figure 2. Topographical AFM images of spin-coated films of (a) RRa-P3HT (rms roughness R_q = 2.2 nm), (b) RR-P3HT (R_q = 1.2 nm), (c) aggRR-P3HT (R_q = 5.4 nm) directly on glass substrate as well as (d) aggRR-P3HT on top of a 30 nm thin polycrystalline Si layer (R_q = 6.3 nm). For (c) and (d), 2% ethyl acetate was mixed with the chloroform solution to enhance the aggregation of the RR-P3HT molecules.

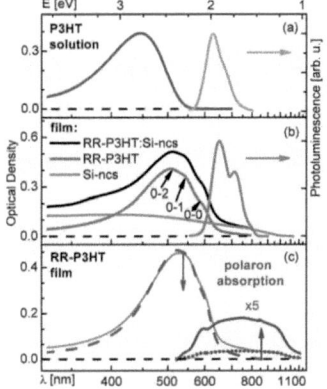

Figure 3. Absorbance and photoluminescence spectra of P3HT in (a) dilute chloroform solution and (b) of RR-P3HT film. The absorbances of Si-ncs and RR-P3HT:Si-ncs (5:1) BHJ as film are also shown. (c) Absorbance spectra of RR-P3HT (blue dashes), oxidized RR-P3HT film after adding FeCl$_3$ (solid gray curve) and P3HT film polaron absorption (brown dots and brown solid curve).

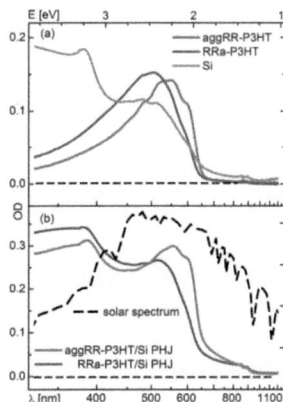

Figure 4. (a) Absorbance spectra of neat RRa-P3HT, RR-P3HT + ethyl acetate (aggRR-P3HT), and neat poly-Si and (b) the corresponding P3HT/poly-Si PHJs samples compared with the solar spectrum.

boiling point as the nonideal solvent ethyl acetate (77 °C). The already partially aggregated polymer falls out of solution before the main solvent is fully dissipated and before the film has fully dried, leading to enhanced aggregation evident in the fine structure of the AFM data. Comparison of Figure 2c and Figure 2d further reveals that when spin-coating aggRR-P3HT directly on poly-Si instead of glass, the fine structure indicating the aggregation is maintained.

3.2. Optical Characterization. The amount of aggregation present in a P3HT sample manifests itself not only in the AFM characteristics but also in the optical spectra. Figure 3 shows the absorbance (OD) and photoluminescence (PL) spectra of RR-P3HT in the dilute chloroform solution and for thin film samples used in this work. In dilute solution with a good solvent such as chloroform, RR-P3HT is known to adopt a random coil conformation with a distribution of short conjugation lengths. This results in a structureless absorbance that has its maximum at 446 nm and in a red-shifted more structured PL with peaks at 627 and 688 nm (Figure 3a). The PL originates from the longest polymer segments after relaxation of the initial photoexcitation. In film, the RR-P3HT chains can planarize to form weakly interacting H-aggregates that are embedded in a matrix of amorphous coiled P3HT chains. The resulting absorption thus consists of a superposition of absorption by coiled chains and absorption by planar, highly conjugated and aggregated chains.[21] Consequently, the absorption is shifted to longer wavelengths, is broadened, and shows vibronic peaks at 518, 558, and 608 nm (Figure 3b) due to the 0−2, 0−1, and 0−0 transitions, respectively, in agreement with literature data.[21] The corresponding PL also contains vibronic structure with peaks at 650, 712, and 800 nm (mainly the C=C symmetric stretching mode, 1452 cm^{-1}).

By considering the intensity of absorption between 400 and 500 nm and by considering the vibrational structure around 600 nm through a modified Franck-Condon analysis as described in refs 21 and 47, it is possible to derive the amount of aggregates present in a film (Figure S2). Figure 4a shows the absorption spectra of the two limiting cases given in our study, that is, RRa-P3HT and aggRR-P3HT. The absorption of poly-Si is also shown for comparison. Analyzing the P3HT absorption in this fashion, with the assumption that the absorption coefficient of aggregated chains is 1.39 times that of coiled chains,[47] yields a percentage of (38 ± 5)% aggregates for the aggRR-P3HT sample and of (24 ± 5)% aggregates for the RRa-P3HT sample (Figure S2). RRa-P3HT possesses less aggregation and a reduced conjugation length compared to RR-P3HT, which was attributed to the adverse steric repulsive interactions between the hexyl side chains and the sulfur.[17,48] Figures 3b and 4b illustrate that when the P3HT is intermixed with Si-ncs or spun on top of poly-Si, the absorption spectrum is given by a superposition of the individual components. Thus, in agreement with the AFM data, spinning a film on top of a silicon substrate does not seem to affect the amount of aggregates formed. Figure 4b further demonstrates the good match of the heterojunction absorption against the solar irradiation spectrum that is essential for efficient solar cells.

In order to study the process of charge generation in these material systems, we do not only need to know how many aggregated or coiled chains are present in the P3HT but also require a spectroscopic signature for charges in P3HT films. P3HT thin films can be chemically oxidized employing a strong oxidant.[25,49] Figure 3c shows the absorbance spectrum of thus oxidized RR-P3HT thin films after dipping into 20 ppm solution of iron(III) chloride ($FeCl_3$) in acetonitrile (CH_3CN) for 1 min followed by rinsing with acetonitrile to remove excess oxidant. From the raw spectrum of the treated film we obtain the P3HT polaron ($P3HT^+$) absorption as follows. The treatment with $FeCl_3$ decreases the known absorbance of the neutral P3HT molecule (blue dashes) and increases absorbance in the range from about 560 to 1150 nm (brown dots). Partial saturation occurs at treatment longer than 2 min. The difference between the blue dashed line scaled to the peak of the gray solid curve in Figure 3c and the gray solid curve reveals the RR-P3HT film polaron absorption (brown dots and brown solid line). It ranges from about 560 to 1150 nm with a characteristic shape and increases again up to the mid-infrared spectral region. The same oxidation was performed for RRa- and aggRR-P3HT, and the results are shown in Figure 5. Comparable but, because of reduced conjugation length, slightly shifted results of the cation absorption are obtained for P3HT in solution where chemical oxidation was introduced via adding pentachloroantimonate (Figure S6, see details in Supporting Information). Similar oxidation experiments as the ones presented here have been reported and yielded very similar spectra.[25,49,50]

The polaron absorption spectrum, that we obtain in the P3HT films (Figure 3c), closely matches the subgap polaron absorption bands obtained by CW photoinduced absorption (PA) measurements as reported in recent publications.[7,11,25,51,52] Therefore, we use our measured polaron absorption spectra for the individual P3HT types for comparison with the transient absorption spectra throughout the whole work.

3.3. Primary Photoexcitations. The thoroughly characterized samples and the systematic variation of their composition allow the study of the excitations formed under illumination. To directly probe the nature of the primary photoexcitations and the inherent photovoltaic conversion processes in hybrid P3HT/Si layers, we performed ultrafast pump—probe (transient absorption, TA) spectroscopy. The thin film samples were designed to enable direct comparison between neat P3HT, neat Si, and the P3HT/Si heterojunction in ultrafast TA spectroscopy at the same experimental conditions by moving the relevant sample regions into the pump—probe region. We chose pump and probe beam diameters at the sample that allow for ensemble averaging over the finely grained morphology of the thin films to mimic their usage as photovoltaic device and to ensure low excitation fluence ($9\,\mu J/cm^2$ at 518 nm) comparable to the solar exposure. The low excitation fluence was also chosen to prevent modulation of the transient signatures due to a thermally induced spectral blue shift, which might occur at high excitation fluences.[53]

Figure 5 shows typical Vis—NIR TA spectra for aggP3HT/poly-Si PHJ, RRa-P3HT/poly-Si PHJ, and RR-P3HT:Si-ncs BHJ (red solid curves) compared with the TA spectra of the corresponding neat polymer film (black solid curves). Spectra were recorded for the full range from 10 ps prior to the pump pulse up to 2 ns after the pump pulse (see Figure S11). Spectra taken at 300 fs, 20 ps, and 1-2 ns pump—probe delay are shown. The spectra are composed of negative ΔOD signals in the spectral region of the P3HT film absorption between 415 and 630 nm because of ground state bleach (GSB) and positive signals at longer wavelengths due to photoinduced absorption (PIA), i.e. excitons and polarons.

In order to directly compare the transient spectra and the kinetics of the charged species of the various neat P3HT with the corresponding P3HT/Si heterojunctions, the *same number of initial photoexcitations* needs to be considered. Small deviations in this number of initial photoexcitations occur in the experiment due to small variations in film quality, film thickness, and thus absorption or due to small changes in pump beam size and pump energy between the various samples. The integral over the GSB area equals the product of the excitation density times the strength of the first electronic transition of P3HT regardless of small variations of Franck-Condon activity due to differing morphologies of P3HT in the various samples.[54] The integral therefore represents a good relative measure of the number of initial photoexcitations. We use the transient spectrum of the aggRR-P3HT film as reference. The TA spectra of the other polymers and PHJs are each scaled by a small factor (0.7—1.3) for the whole data set of 250 time steps and 512 wavelengths. As a result, all spectra yield the same GSB area integral at the earliest usable time delay. We found that any delay time between 0 and 80 fs gives the same result. This procedure allows for considering the same number of initial photoexcitations in the films. Any changes in the optical density (ΔOD in %) that would be seen under reference conditions are then solely inherent to the different nature of the sample.

For each measurement, the pairwise difference between TA spectra of the heterojunction and the neat polymer is calculated (green solid curves). For comparison, the particular P3HT film polaron absorption spectrum (blue dashes) is included. The GSB of aggRR-P3HT and RR-P3HT clearly reveals the characteristic vibronic structure of the P3HT aggregate, in particular at 1-2 ns after excitation, which does not show a spectral shift during the delay times covered in the TA measurements. In Figure 5, we further observe the relative signal ratio between the 0—0 and 0—1 GSB peak to change over time and, most importantly, the GSB at 500 nm to reduce with time. As detailed further below in section 3.6.1 and in the Supporting Information, this is a signature that energy transfer takes place from coiled chains to aggregated chains.

Since the GSB is due to those P3HT chains which are not in the electronic ground state, it directly monitors the total number of photoexcitations such as excitons or polarons still present at a given delay time. The broadband PIA signals with positive ΔOD (compare Figure 5) are adjacent to the GSB and exceed the NIR detection range. We compare the PIA for the hybrid heterojunctions (red solid curves) to the corresponding neat polymer (black solid curves) and the difference spectra for each pair with the chemically obtained P3HT film polaron absorption spectrum. We find that the difference spectra match the P3HT film polaron absorption spectra in the range of about 620—850 nm. The comparison also reveals a region of enhanced PIA through adding Si to P3HT. From this we conclude that combining P3HT with silicon results in the generation of

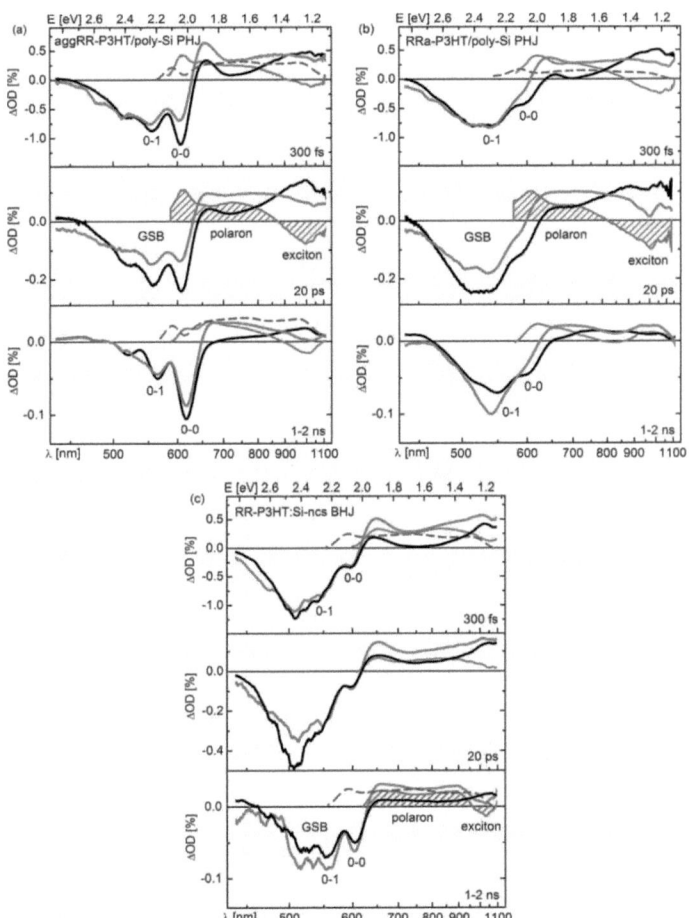

Figure 5. Transient absorption of (a) aggRR-P3HT/poly-Si PHJ, (b) RRa-P3HT/poly-Si PHJ and (c) RR-P3HT:Si-ncs BHJ for 300 fs, 20 ps and 1−2 ns pump−probe delay. The spectra are scaled to the same number of initial photoexcitations. The P3HT polaron absorption (blue dashes) and the difference (green solid curve) between neat P3HT (black solid curve) and P3HT/Si (red solid curve) are compared. Excitation: 518 nm, 9 μJ/cm^2.

additional P3HT cations, i.e. polarons. In the neat P3HT film, this polaron absorption also seems to be present, albeit at a significantly reduced level. From a comparison of the signal magnitude at 660 nm in the neat P3HT film and in the hybrid heterojunction we infer that the polaron yield in the P3HT/Si heterojunctions is more than a factor of 2 higher. Clearly, there must be ultrafast electron transfer from P3HT to Si in PHJ and BHJ morphologies. Consequently, Si is a promising electron acceptor for hybrid photovoltaic devices, in accordance with results by LESR.[9]

We now focus on the PIA spectra in the range of 900−1100 nm. In this range, the difference spectrum, given by the green solid curve in Figure 5, does not match the polaron absorption spectrum (blue dots). In agreement with previous investigations,[16,25,29,30] we assign the photoinduced absorption in this spectral range to absorption by singlet excitons. To support this assignment, we performed TA measurements of RR-P3HT in various dilute solutions, where the intermolecular distance is high. We found that under these conditions, P3HT cation formation does not take place (Figure S7, see details in Supporting Information).

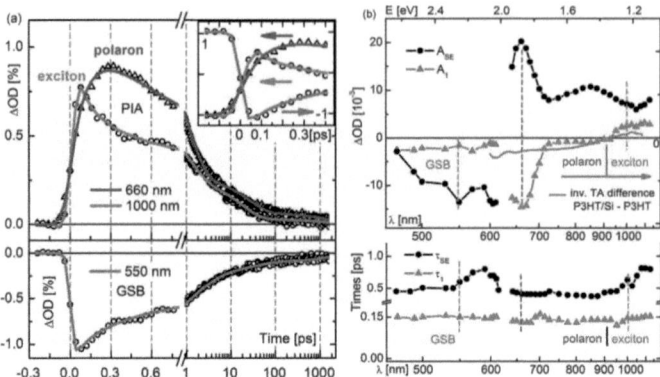

Figure 6. (a) Experimental TA signals (open symbols) as function of the pump−probe delay with corresponding fits (solid curves) of GSB (green), singlet exciton (red), and polaron (blue) for the aggRR-P3HT/poly-Si PHJ. The fit parameters are listed in Table 2. (b) PIA band single-channel fit amplitudes and time constants for the 140 fs component (red triangles) and the 1 ps component (black dots). The dashed lines are the signals taken for (a). The band assignment (inverted green curve) from Figure 5a agrees with the separation of polaron and exciton signatures via the fit amplitudes of the 140 fs component.

The singlet exciton leads to stimulated emission and undergoes intersystem crossing on the nanosecond time scale to a triplet state with a characteristic transition at 1.49 eV (830 nm). The energetic position of the P3HT triplet transition confirms and adds to recent experimental and theoretical investigations.[16,55] In addition to confirming the assignment of the singlet and triplet excitons, this measurement also implies that excitation dissociation does not occur efficiently on a single polymer chain.

Moreover, the difference spectrum between the hybrid system and the neat P3HT in Figure 5 shows a reduced PIA in the range of 900−1150 nm, which suggests that adding silicon to P3HT rapidly reduces the number of singlet excitons. The time evolution of the TA signals will be analyzed in detail further below. No transient signatures of stimulated emission (SE) or a triplet state are detected in any of the film measurements. This observation and the concomitant increase of polaron absorption and reduction of exciton absorption when P3HT is combined with silicon suggest the dissociation of singlet excitons as a path for polaron formation. We therefore conclude that the enhancement of polaron yield in P3HT by adding Si is due to an ultrafast electron transfer from P3HT (electron donor) to Si (electron acceptor) for all morphologies, namely PHJ and BHJ. We note that no transient signatures could be detected for the neat Si films or for excitation of the P3HT/Si heterojunctions with a pump wavelength of 720 nm, which is outside the P3HT absorption (Figures 3b and 4a). This shows that the observed effect of enhanced polaron absorption and reduced exciton absorption is due to electron transfer from excited P3HT to Si and not due to optical effects such as a transient change of index of refraction in Si.

So far we have assigned the various features of the TA spectra to photoexcitations, and the general observations pertained to all sample structures. We now consider the quantitative differences that arise between the various samples. In agreement with previous work,[16,29,33] we find some polaron formation to occur also for the neat P3HT films. Taking the signal magnitude around 660 nm as a measure for the amount of polaron formation, we find the initial yield to decrease in the order aggRR-P3HT, RR-P3HT, and RRa-P3HT. This is also the order in which the amount of aggregated chains in the film decreases (Figure 2). Therefore, the obtained clear discrimination between the TA spectra of the three types of P3HT matches the distinct differences in the corresponding optical and structural properties. We thus associate a higher polaron yield with an enhanced degree of aggregation and extended conjugation, enabling highly delocalized excitations and charge carriers with high mobility.[24,25]

This effect of the P3HT structural order on the polaron yield is also manifested in the hybrid heterojunctions. Comparison of the polaron absorption signal (at 660 nm; see red curve in Figure 5) indicates about a factor of 2 more efficient initial polaron formation for aggRR-P3HT/poly-Si PHJ compared to RRa-P3HT/poly-Si PHJ. We attribute this finding to the higher degree of conjugation, pronounced exciton delocalization, and an increased mobility of charge carriers which enable a more efficient charge transfer. A possibly more favorable free energy of charge generation through improved band alignment for ordered P3HT chains remains to be investigated.[56] Recent results state a rise only of the HOMO energy level and no change of the LUMO level through chain ordering.[57] For a possible photovoltaic application, we compared RR-P3HT:Si-ncs BHJs with aggRR-P3HT/poly-Si PHJs. The polaron yield is higher and their lifetime is slightly longer in aggRR-P3HT/poly-Si PHJ than in RR-P3HT:Si-ncs BHJ as becomes evident through comparison of the TA spectra in Figure 5a,c. The former can be attributed to the intermediate P3HT type used in the BHJ. The latter is assigned to Si dangling bond defects in Si-ncs which act as recombination centers. To address this, we are currently working on postgrowth treatments, e.g., HF etching, vacuum annealing, and surface functionalization, which improve the surface and defect properties.[58] Moreover, interrupted percolation paths in the current BHJ morphology can also lead to enhanced recombination and limit the efficiency, as not the entire amount of photoinduced charge carriers can move to the electrodes. To overcome this, the fabrication of a defined BHJ morphology via nanoimprinting is a promising approach.

In summary, our TA measurements point out that polymer structural order plays a significant role in hybrid solar cells, though Si can significantly enhance the initial polaron yield even in the disordered RRa-P3HT (Figure 5b). Below, it will be investigated whether bound polaron pairs or mobile charge carriers are formed in the P3HT/Si heterojunctions.

3.4. Temporal Evolution of the Charge Generation Process.

Having identified the features of GSB (about 415−620 nm), polaron absorption (about 620−900 nm), and singlet-exciton absorption (900−1150 nm), we now consider their inherent kinetics in a quantitative fashion. Figure 6a shows the evolution of the TA signals for the GSB (at 550 nm), the polaron absorption (at 660 nm), and the singlet exciton absorption (at 1000 nm) of aggRR-P3HT/poly-Si PHJ on a time scale up to 2 ns with a time resolution of 40 fs. From Figure 6a, a few observations can be made immediately. First, the GSB and the singlet exciton absorption reach their maximum signal with a rise time of 40 fs (see also the inset), i.e., within the experimental resolution. This observation is important, as it implies that the exciton is formed *directly upon photoexcitation*. Second, the polaron absorption signal shows a *delayed* rise. It reaches its maximum at about 300 fs. This rise of the polaron signal is matched by a corresponding initial decay of the singlet exciton absorption signal. This suggests that the singlet exciton, formed by absorption, decays to form polarons. Third, the ultrafast time evolution of the GSB and the singlet exciton absorption signal is very similar. This can be readily understood by an additional ultrafast nonradiative decay mechanism of at least some of the excited P3HT molecules. A minimal rate model that allows relaxation of the exciton into the polaron by dissociation and to the P3HT ground state by nonradiative decay directly renders the result that the yield of each channel is given by the ratio of the individual rate to the sum of both.[59] The spectroscopic signal is additionally weighted by the respective extinction coefficients. The exciton serves as a reservoir, and the same femtosecond kinetics is observed for the decay of the exciton signal and the recovery of the GSB.[59] Nonradiative electronic decay on the femtosecond time scale is now widely reported for a large variety of molecular systems[60−64] and believed to be frequently mediated by conical intersections.[65,66] Whether a conical intersection is also responsible for the observed ultrafast nonradiative decay in P3HT films has to be clarified in the future. It has recently been established in conjugated polymers that exciton localization occurs in tens of femtoseconds and leads to nonemissive states.[67,68] Single-molecule spectroscopy has correlated the ultrafast relaxation to aggregated regions of the polymer.[69]

To further substantiate the conclusion that the polarons are formed from the excitons, we have fitted the decay of the transient data. Already a visual interpretation of the kinetic traces shows that there is the ultrafast signal change as discussed above and an additional slower component. It is possible to model the decay of the GSB and the singlet exciton absorption as a stretched exponential (SE) curve. A stretched exponential decay is expected for films that possess an ensemble of ordered and disordered regions and a correspondingly broad distribution of decay times.[70,71] As we are particularly interested in the initial signal changes, we treat the ultrafast component separately by fitting the GSB and the singlet exciton absorption according to the function

$$\Delta OD = A_1 \exp(-t/\tau_1) + A_{SE} \exp(-t/\tau_{SE})^\beta + \text{const} \quad (1)$$

We also use eq 1 to fit the polaron absorption signal. The polaron decay will also be characterized by a distribution of

Table 2. Fit Parameters: Amplitudes and Time Constants for the Transient Species in Figure 6

parameter	polaron (660 nm)	exciton (1000 nm)	GSB (550 nm)
A_1	-14.6×10^{-3}	2.6×10^{-3}	-1.6×10^{-3}
τ_1 (ps)	0.14	0.14	0.14
A_{SE}	20.2×10^{-3}	7.1×10^{-3}	-13.6×10^{-3}
τ_{SE} (ps)	0.41	0.63	0.6

relaxation times, because of various on-chain and interchain recombination paths. Therefore, the use of a stretched exponential fit for the decay kinetics is justified.[71] The fit curves obtained are indicated as solid colored curves in Figure 6a. The fit parameters A_1, A_{SE}, τ_1, τ_{SE} are listed in Table 2 for the ΔOD signals at 660 nm (polaron), 1000 nm (exciton), and 550 nm (GSB). The exponent β was found to be 0.5 in all cases. Figure 6b illustrates how these fit parameters vary as a function of probe wavelength across the entire detected spectral range.

We find that the singlet exciton absorption and the GSB both decay with a similar first ultrafast time constant of 140 fs, followed by a slower decay, for which the combination of $\tau_{SE} \approx 0.6$ ps and $\beta = 0.5$ yields an average decay time $\bar{\tau}$ of about 1.2 ps. We find the *same* time constant of 140 fs for the delayed buildup of the polaron signal, followed by a decay characterized by an average decay time $\bar{\tau}$ of about 0.8 ps. The fact that the rise of the polaron population is correlated with a simultaneous decay of the exciton population is strong evidence that the polarons are created through the dissociation of singlet excitons.[72]

From the overview of the fit parameters in Figure 6b, we see that the time constants found in our fits stay rather constant over the entire spectral range of the individual transient species. Moreover, the fastest time constant τ_1 is present over the entire probe spectral range. This implies that the chosen model describes the intrinsic dynamics properly. The variations of the τ_{SE} decay time between the transient species can be attributed to different recombination processes for polarons and excitons as well as energy transfer processes between coils and aggregates suggested above and demonstrated in more detail below. The amplitudes for the ultrafast 140 fs component (A_1) and the slower component (A_{SE}) have the same sign in the spectral region describing the singlet exciton absorption, that is, from 900 to 1150 nm and beyond. For shorter probe wavelengths, where the polaron absorption is probed, the amplitude A_1 for the 140 fs component changes sign, as it no longer describes an absorption decay but rather the delayed rise of the polaron absorption. This evident correlation between singlet exciton decay and polaron rise further strengthens our interpretation. Similar kinetics are observed for pure P3HT, RRa-P3HT/poly-Si PHJ, and RR-P3HT:Si-ncs BHJ, independent of pump polarization orientation. At 6 ns, the GSB shows 5% of its initial signal magnitude that we assign to long-lived photoexcitations, which can readily be harvested in a solar cell. At this point we want to emphasize that the samples were optimized for ultrafast TA measurements and not for optimum solar cell performance, where reduced recombination can be achieved via various methods as detailed in the introduction and the conclusion sections.

For an absorbed number of photons of 2.6×10^9 per pulse at an excitation of 9 μJ/cm^2 at 518 nm, we obtain an excited state areal density of 2.4×10^{13} cm^{-2} (Table 3, see details in Supporting Information). The polaron cross section of $(3.4 \pm 2) \times 10^{-16}$ cm^2, the polaron molar extinction coefficient of

Table 3. Parameters of the P3HT/Si Heterojunction with 9 μJ/cm² Excitation at 518 nm

parameter	
absorbed photons	2.6×10^9/pulse
areal density of excitations (cm^{-2})	2.4×10^{13}
polaron cross section (cm^2)	$(3.4 \pm 2) \times 10^{-16}$
initial charge density (cm^{-3})	6×10^{18}

Table 4. Initial Quantum Yields of Charges in Neat P3HT and aggRR-P3HT/poly-Si PHJ

Sample	P3HT$^+$	P3HT$^-$	P3HT + E	Si$^-$
aggP3HT	0.17	0.17	0.83	0
aggP3HT/Si	0.38	0.13	0.62	0.25

$(4 \pm 1) \times 10^4$ L mol^{-1} cm^{-1}, and the initial charge density of 6×10^{18} cm^{-3} obtained from our measurement series are similar to previous investigations of RR-P3HT:PCBM BHJ at comparable excitation fluence.[16,30,73]

With the interpretation that photoexcitation generates excitons that subsequently decay into polarons, it is possible to estimate the maximum yield of polarons formed initially by considering the relative magnitudes of the TA signals at 300 fs given in Figure 5. Details of the calculations can be found in the Supporting Information. While we estimate a maximum quantum yield of 17% for the formation of the P3HT$^+$ polaron in a neat film of aggRR-P3HT, this value raises to 38% in the planar heterojunction (aggRR-P3HT/poly-Si PHJ) in combination with a maximum Si$^-$ yield of 25% (Table 4). We note that in the heterojunction device, the P3HT layer covering the silicon has a film thickness of only 40 nm so that excitons are created close to the donor−acceptor interface. The significant enhancement of exciton dissociation in the presence of silicon implies that silicon performs very well as an electron accepting material. The charge yield points to efficient electron transfer and a spatial exciton delocalization in ordered P3HT of about 10 nm, in agreement with previous investigations.[74,75]

The obtained polaron yields are similar to RR-P3HT:PCBM studied previously,[29] which raises the hope for aggRR-P3HT/Si heterojunctions to achieve comparable power conversion efficiencies as existing and even commercially available thin film solar cells based on RR-P3HT:PCBM BHJs.[32]

3.5. Charge Recombination Processes. So far we have substantiated the discussion of the yield and the time scale of the exciton dissociation into polarons. Now we address the issue of whether the positive and negative polarons formed are still Coulomb-bound as a geminate pair or whether they are free charge carriers. This question is crucial for the efficient operation of photovoltaic devices. There is no obvious reason why the TA spectra of bound polarons should be very different from those of free polarons. We are therefore not able to distinguish between bound and free polarons on the basis of the TA spectra. However, it is possible to distinguish the two species by considering their recombination kinetics. We expect a pair of geminately bound positive and negative polarons to recombine (radiatively or nonradiatively) with each other, i.e., monomolecularly. Their decay should therefore not be affected by the overall number density of bound polaron pairs that are formed upon photoexcitation. Further, the number density of bound polaron pairs formed should not impact on the singlet exciton population, thus leaving the decay kinetics of the singlet exciton absorption or the GSB unaltered. In contrast, if exciton dissociation results in the generation of free, i.e., nongeminate, positive and negative polarons, then these charges can only recombine when they meet each other, i.e., by bimolecular charge−charge annihilation. The probability of meeting the oppositely charged polaron thereby increases with the number density of polarons formed.[16,73] Further, singlet excitons can recombine through quenching by free polarons, and the probability for this exciton-charge annihilation also increases with the number density of free polarons.[16,76,77] Thus, in summary, for geminately bound polarons we expect the decay transients of polaron absorption, singlet exciton absorption, and GSB to be independent of excitation fluence, while we expect accelerated decays with increasing excitation fluence for free polarons generated upon singlet exciton dissociation.

To study the role of delocalization and the nature of the recombination processes in P3HT/Si heterojunctions, we record Vis−NIR TA spectra up to 2 ns delay with increased excitation fluences from 4 to 60 μJ/cm². The absorption signals of the corresponding relevant transient species in aggRR-P3HT/poly-Si PHJ (Figure 7a), RRa-P3HT/poly-Si PHJ (Figure 7b), and RR-P3HT:Si-ncs BHJ (Figure S8) are normalized against their individual initial maximum. This allows the study of the recombination rate as a function of excitation fluence. For the RRa-P3HT/poly-Si PHJ sample we find the decay kinetics to be independent of excitation fluence. Consequently, the polarons formed in a PHJ of silicon with RRa-P3HT are predominantly Coulomb-bound. In contrast, we find enhanced decay rates with increasing excitation fluence for aggRR-P3HT/poly-Si PHJ, suggesting the predominant formation of free polarons. The same observation is made for RR-P3HT:Si-ncs BHJs (Figure S8). In Figure 7a the enhanced decay rate leads to an apparent shift of the maximum of polaron absorption to earlier times, covering the delayed polaron rise.

Consequently, in the aggRR-P3HT/poly-Si PHJ and the RR-P3HT:Si-ncs BHJ photoinduced ultrafast generation of free charges is obtained, which opens the route for efficient charge extraction from the active layer in the hybrid devices. In fact, it was recently shown that the competition between extraction and bimolecular recombination of mobile charges determines the dependence of the photocurrent on the applied bias and therefore the fill factor in RR-P3HT:PCBM BHJ devices.[78,79]

It is worthwhile to briefly reflect on these results. For the RRa-P3HT/poly-Si PHJ, photoexcitation near the hybrid interface results in excitons that are mostly localized on coiled chains. An electron is then transferred to the poly-Si with a time constant of 140 fs, leaving behind a P3HT$^+$ cation, i.e., a positive polaron. In silicon, the electron can be expected to be well delocalized because of the high dielectric screening. However, on the coiled P3HT chain, the conjugation length is low and the energetic disorder is high. As a result, the positive polaron is localized in the sense that its coherence length and its mobility are low. It seems that the presence of a more "pointlike" and moreover "immobile" positive charge on the P3HT prevents the formation of free polarons. In a certain way, the situation is comparable to that of a point charge in front of a metal that feels an attractive force. In contrast for the aggRR-P3HT/poly-Si PHJ, a significant fraction of the excitations are created on planar, aggregated chains that are characterized by a high conjugation length and low energetic disorder. After charge transfer, the delocalized electron in the silicon is thus interacting with a positive polaron that not only is

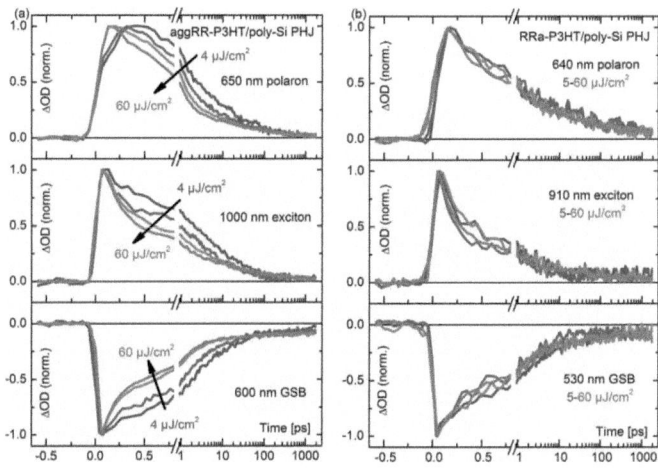

Figure 7. (a) Decay of TA signals for increased excitation fluences (4, 10, 40, and 60 $\mu J/cm^2$ at 518 nm) reveals bimolecular nongeminate recombination in the case of aggRR-P3HT/poly-Si PHJ. (b) Decay of TA signals for increased excitation fluences (5, 20, 40, and 60 $\mu J/cm^2$ at 518 nm) reveals monomolecular geminate recombination in the case of RRa-P3HT/poly-Si PHJ.

comparatively delocalized but also has a high initial mobility due to the low disorder. Such a polaron might move away from the hybrid interface, for example, by spectral diffusion to slightly longer conjugated segments, thereby overcoming the weak Coulomb attraction to the delocalized electron in the silicon and thus leading to the formation of a free pair of positive and negative charge. This finding is also supported by anisotropy measurements (Figure S10). It seems that a key issue in exciton dissociation is the delocalization, the dielectric screening, and the mobility of both the electron *and* the hole. Thus, a significant implication of the polymer structural order for hybrid and organic solar cells becomes evident and can be understood on a microscopic level.

Our interpretation of the results is based on the concept of singlet exciton dissociation into bound or free polaron pairs that we support by Figures 5 and 6. From the excitation fluence of 4 $\mu J/cm^2$, where bimolecular annihilation sets in, we derive a mean excitation spacing of more than 6 nm in aggRR-P3HT and RR-P3HT, assuming an isotropic distribution of photoexcitations within the pump−probe volume. This finding reveals that despite the 6 nm (∼15 thiophene repeating units) separation between the initial photoexcitations, the subsequent bimolecular interactions due to the generated mobile polarons can still take place pointing to high charge carrier mobility and rather spatially delocalized excitations, in agreement with other TA measurements for conjugated polymers,[29,74] and supporting the degree of delocalization indicated from our quantum yield calculations mentioned above.

It was previously suggested that polarons in conjugated polymers may also be generated from higher excitonic states accessed by sequential excitation or by exciton−exciton annihilation.[33,80−82] These delocalized "hot exciton" charge-transfer states are supposed to exhibit a higher dissociation probability via enhanced electron−hole separation and charge mobility.[75,83,84] These additional processes cannot, however, be the dominant polaron formation pathway in the present work for several reasons. First, the polaron formation happens with a 140 fs time, which is too fast for bimolecular annihilation processes.[16,77,82] Second, sequential excitations during the pump pulse come into play only at very high excitation fluences starting at 100−400 $\mu J/cm^2$.[81,82] Third, the polarons show the same intensity dependence as the excitons (Figure S9) and are thus generated from singlet excitons.

3.6. Variation of Excitation Wavelength.

3.6.1. Energy Transfer. We have seen that the conformation of the polymer chain has a major impact on the nature of the photogenerated charges. The number of excitations created on coiled chains or on aggregated chains is determined not only by the choice of P3HT regioregularity and solvent but also by the choice of excitation wavelength. Figure 8 shows ultrafast UV−Vis TA spectra of neat RR-P3HT thin films at 60 fs, 300 fs, 13 ps, and 140 ps with excitation at 450 nm (blue solid curve) and 600 nm (red solid curve). The spectra are normalized against the GSB so that signal changes solely inherent to the variation of the pump wavelength can be studied. The transient absorption spectra contain signatures of GSB (425−625 nm) and of polaron absorption (>625 nm). The inverted RR-P3HT film absorption spectrum (OD, green solid curve) is scaled to the RR-P3HT GSB peaks. The calculated difference spectra between the GSB and the inverted thin film absorption are shown as dashed curves for both excitation wavelengths (TA − OD, dashed). These difference spectra are then compared to the scaled absorption spectrum of RR-P3HT in dilute chloroform solution (OD, cyan solid line).

At 60 fs, the GSB of the RR-P3HT film is significantly broader when excited at 450 nm than for excitation at 600 nm. Immediately after excitation with 600 nm, the GSB of the film lacks the spectral components equivalent to the absorption spectrum in dilute chloroform solution. In dilute chloroform solution, polymers form coils as the chains curl up. Thus, at 60 fs after excitation, only planar

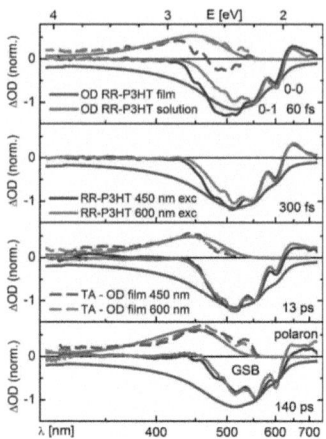

Figure 8. UV–Vis TA spectra of RR-P3HT thin film at 60 fs, 300 fs, 13 ps, and 140 ps with excitation at 450 nm (solid blue curve) and at 600 nm (solid red curve). The inverted absorption spectrum of RR-P3HT thin film (OD, green solid curve) is scaled to the RR-P3HT GSB peaks to extract the differences (TA − OD, corresponding dashed curves) between the transient spectra and the film absorption, which are compared to the absorption spectrum of RR-P3HT in dilute chloroform solution (OD, cyan solid curve).

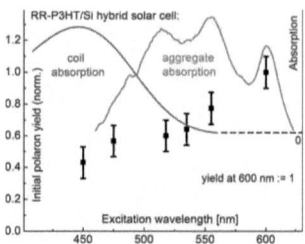

Figure 9. Dependence of initial polaron yield (black squares) on excitation wavelength for RR-P3HT/Si heterojunctions. Selective excitation of coiled (blue curve) vs aggregated (red curve) RR-P3HT domains reveals greater than a factor of 2 more efficient charge separation if exciting directly the aggregated RR-P3HT domains.

aggregated chains are excited by light with 600 nm, while both coiled and aggregated chains are excited by light with 450 nm. This distinct difference between the spectra obtained for excitation with 450 nm and with 600 nm stays visible up to 13 ps. At 13 ps, the film GSB coincides for both pump wavelengths. The difference between the GSB at 13 ps and the inverted film absorption spectrum corresponds to the absorption spectrum of P3HT coils except for some weak low energy tail around 480 nm.

At 140 ps after excitation, the GSB shows even clearer vibrational structure and the difference between GSB and inverted film absorption spectrum reveals spectral parts missing in the GSB which are even beyond the coil absorption spectrum. This indicates further slow energy migration within the aggregates toward more planar and extended conjugated segments, e.g., via torsional relaxation or excitation energy transfer (EET).[85,86] Thus, the GSB beyond 13 ps reveals the absorption spectrum due to RR-P3HT aggregates only, which is taken from the 140 ps case for Figure 9. It can be seen that aggregate absorption ranges from 460 to 625 nm.

For excitation at 600 nm, the GSB at 60 fs and the GSB at 140 ps are identical except for the changes in the 0−0 peak intensity already discussed in the context of Figure 5. In contrast, for excitation at 450 nm, the GSB loses the higher energy contributions that are attributed to coiled chains. We attribute this behavior to an energy transfer from unordered (coils) to ordered domains (aggregates, planar segments) in RR-P3HT films, as also indicated in the TA measurements of the PHJs and the BHJ above. Parts a and b of Figure 3 show that the PL of RR-P3HT in dilute solution (coils) overlaps with the absorption spectrum of RR-P3HT thin films, which is a prerequisite for efficient electronic energy transfer.[68,70] With excitation of the RR-P3HT film at 450 nm, both coils and aggregates are addressed; however, the coils undergo downhill energy transfer to the aggregated regions. An analysis of the ratio between the GSB intensity at 480 nm for excitation at 450 and at 600 nm reveals a forward $1/e$ energy transfer time of about 3 ps (Figure S5, see more details in the Supporting Information). It is interesting to note that this is the same time constant as observed for the energy transfer from glassy phase to a planarized phase in poly(9,9-dioctylfluorene).[37]

3.6.2. Role of Excess Energy. Figure 8 allows us to distinguish the absorption range of coils and aggregates. For excitation wavelengths above 500 nm one predominantly addresses the aggregated RR-P3HT regions. For excitation above 550 nm one exclusively addresses the aggregated RR-P3HT regions. Furthermore, with excitation wavelengths below 500 nm one predominantly addresses the coiled RR-P3HT regions. We can therefore selectively excite ordered or disordered regions of P3HT films. Variation of the excitation wavelength thus provides an alternative approach to control whether coiled or aggregated chains are initially excited. We use this to further corroborate our results obtained on the morphology dependence of the polaron yield. Figure 9 shows the initial polaron yield at about 300 fs as a function of the excitation wavelength at the hybrid heterojunction with the intermediately aggregated RR-P3HT. For comparison, the absorption spectra of coiled and of aggregated P3HT chains (taken as the GSB at 140 ps from Figure 8) are also shown.

For this experiment, the pump pulse was adjusted to central wavelengths of 450, 475, 518, 535, 555, and 600 nm and pulse durations of about 15 fs with the same experimental pump−probe conditions. For directly comparing the transient spectra of hybrid RR-P3HT/Si heterojunctions excited at different wavelengths, the transient spectra were scaled according to the individual initial P3HT GSB (internal standard). In this case, comparing the initial polaron absorption magnitudes of the various TA spectra can reveal changes in polaron yield solely inherent to the different excitation wavelengths while considering the same number of initial photoexcitations in the P3HT/Si heterojunction.

Figure 9 reveals an enhancement of more than a factor of 2 of initial polaron yield by increasing the excitation wavelength from 450 to 600 nm. The trend was obtained independently on the device structure, i.e., for PHJ and BHJ geometries. The error bars result from multiple measurements of several hybrid RR-P3HT/Si samples under the nominally same experimental conditions.

The increase by a factor of 2 matches the enhancement of the polaron yield which was observed in Figure 5a,b by comparing highly aggregated aggRR-P3HT/poly-Si PHJ with RRa-P3HT/poly-Si PHJ. In the latter case, P3HT coils were predominant. Our findings are an extension of previous TA investigations where the degree of aggregation was varied via thermal annealing.[19] In conclusion, we record the same factor of 2 more efficient charge generation in aggregated polymer-based hybrid heterojunctions compared to the unaggregated version by two different and independent methods: (i) by using different P3HT configurations and therefore solely changing the structural order in the P3HT film (Figure 5) and (ii) by solely changing the excitation wavelength and therefore selectively exciting defined P3HT regions (Figure 9). If excess photon energy was necessary for the exciton dissociation process, we would expect a high initial polaron yield for excitation at 450 nm and a lower polaron yield for 600 nm excitation. The fact that we observe exactly the opposite tendency clarifies that excess photon energy is not required; however, structural order is essential. Whether this order enhances charge separation by increasing the initial charge carrier mobility or by improving the overall energetics or by both remains an intriguing question for further research.

4. CONCLUSIONS AND IMPLICATIONS FOR HYBRID AND ORGANIC PHOTOVOLTAIC DEVICES

We have comprehensively studied the nature of primary photoexcitations and their inherent dynamics in neat P3HT and in hybrid P3HT/Si thin films by ultrabroadband (UV–Vis–NIR) transient absorption (TA) spectroscopy with 40 fs time resolution and varied excitation wavelength. Hybrid heterojunctions with 30 nm thin polycrystalline Si layers or Si nanocrystals were processed with P3HT of varied polymer structural order and film geometry. The spatial and optical properties of planar and bulk heterojunctions show that Si does not change the P3HT structure and leads to a broad film absorption range from the UV to 1100 nm needed for efficient light-harvesting.

Scheme 1 summarizes the primary photoinduced processes in hybrid P3HT/Si thin film heterojunctions. In the TA experiments, we can identify the transient signatures of P3HT polarons (620–900 nm) and singlet excitons (900–1150 nm). Our measurements reveal singlet excitons in P3HT as primary photoexcitation with a subsequent ultrafast electron transfer from P3HT to Si as inherent photovoltaic conversion process for all employed hybrid heterojunctions. The addition of Si to RR-P3HT or RRa-P3HT significantly enhances the polaron yield in the active layer

These experiments show that silicon is a particularly favorable electron acceptor because of the highly efficient charge delocalization. Moreover, the higher dielectric constant, compared to the state-of-the-art electron acceptor PCBM, allows for an improved screening of the electron, preventing back transfer. Besides these advantages compared to organic semiconductors, Si is abundantly available and offers the possibility of selective surface modifications and thus interface engineering.

In neat P3HT and in P3HT/Si heterojunctions, we reveal a *delayed* polaron formation compared to singlet excitons, which appear within the experimental time resolution of 40 fs. The population of polarons has a maximum at about 300 fs after excitation. Thus, charge generation is probed in *real time*, revealing a 140 fs rise time for polarons, which is found to correlate with the initial 140 fs decay of the singlet excitons in P3HT. We conclude that the

Scheme 1. Primary Photoinduced Processes in Hybrid P3HT/Si Thin Film Heterojunctions

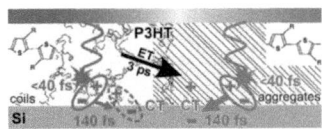

correlated decay of the exciton population and the rise of the polaron population indicate polaron formation via singlet exciton dissociation. In particular we stress that there is no significant polaron population immediately after excitation, i.e., after 40 fs. The ultrafast charge transfer (CT) process implies a strong exchange integral of the excited state orbitals of electron donor and acceptor.

This result demonstrates that the observation of ultrafast charge carrier generation is not in contradiction to the initial formation and subsequent dissociation of a singlet exciton, in contrast to recent suggestions made for the mechanism of charge carrier generation in the blend of the low band gap polymer PCDTBT with PCBM.[31] Their argument essentially pertains to the fact that free charges can be observed as fast as 100 fs after photoexcitation. Similar interpretations for a MDMO-PPV: PCBM blend have been made earlier.[28] This issue, whether light absorption immediately creates free charge carriers *or* excitons, is central to the understanding and optimization of photoconversion in organic and hybrid solar cells. A problem in resolving this issue pertains to the experimental limits of time resolution and spectral range that are available for optical probing. Here we have developed a novel ultrabroadband gap-free TA spectroscopy setup with a time resolution of 40 fs over the entire spectral range from 415 to 1150 nm. This allows us to monitor *both* the decay of the primary excitation and its evolution into a charge pair state.

We have investigated the difference in the polaron formation process for RRa-P3HT, where only a small part of the film is composed of aggregates and for RR-P3HT containing an increased fraction of aggregates. TA measurements with varied excitation fluence clarify that in hybrid P3HT/Si heterojunctions with aggregated P3HT exciton dissociation predominantly leads to free charge carriers, which can in principle be extracted as photocurrent. This is evident from the fact that in RR-P3HT/Si planar as well as in bulk heterojunctions we observe an increasing recombination rate of polarons with increasing excitation fluence, indicating bimolecular nongeminate recombination of charges outside the Coulombic capture radius. These recombination rates indicate that besides the primary photoconversion processes, the charge transport and extraction at the electrodes are crucial issues toward more efficient hybrid photovoltaic devices. Although we have already performed initial systematic studies on the reduction of Si dangling bond defects in the Si-ncs,[58] which act as recombination centers, the charge transport and extraction need to be further addressed in future work. However, power conversion efficiencies of 1% have been shown recently with the first P3HT:Si-ncs blends.[8] Combined with an optimized charge transport and extraction at the electrodes, their efficiency is expected to increase.

In contrast, for RRa-P3HT/Si, where there is initially a higher proportion of excitons on coiled P3HT chains, the decay rate is independent of the excitation fluence, suggesting monomolecular geminate recombination of bound carriers, making it less suitable for photovoltaics from the photophysical perspective. Geminate recombination was recently also found for polymer−polymer blends and seems to be one of the main obstacles to be overcome for photovoltaic applications.[77,87] We attribute this difference between RRa- and RR-P3HT to a more localized hole in RRa-P3HT leading to localized charge carriers in RRa-P3HT/Si as opposed to highly delocalized charge carriers in RR-P3HT/Si revealing that high polymer structural order is a necessity for free charge generation in P3HT/Si.

Quantitatively, we can demonstrate that using aggregated P3HT leads to a *factor of 2 higher polaron yield* compared to employing disordered P3HT in photovoltaics, by two different and independent methods: (i) by using different P3HT configurations and therefore solely changing the structural order in the P3HT film and (ii) by solely changing the excitation wavelength and therefore selectively exciting defined P3HT regions. Combined with results from a modified Franck−Condon analysis, we find that the polaron yield in P3HT/Si increases disproportionally with increasing degree of aggregation in P3HT. Moreover, we find that supplying excess energy does *not* assist the charge carrier separation, whereas our results indicate that ultrafast generation of free charges is *more* dependent on polymer structural order. We argue that the larger conjugation length, low energetic disorder, and the concomitant higher initial charge carrier mobility in the planar aggregated P3HT compared to the short conjugation length in coiled P3HT favor the dissociation process into free charge carriers. Additionally, we observed that downhill energy transfer (ET) from coiled to aggregated chains takes place with a time constant of 3 ps.

For this reason, purely organic and hybrid photovoltaic devices using P3HT should employ highly aggregated P3HT. The loss of high-energy polymer absorption can be compensated by stacking heterojunctions in tandem or even multiple solar cells using conjugated polymers with different band gaps.[88,89] Detailed investigations of the optimum Si-nc band gap for charge transfer remains for future research. The present ultrafast spectroscopic studies combined with ongoing P3HT/Si-ncs device optimization in terms of surface and defect properties of the Si-ncs as well as of the film morphology raise the hope to realize efficient P3HT/Si photovoltaic devices.

■ ASSOCIATED CONTENT

ⓈSupporting Information. Basic optical and structural characterization of P3HT: 3D AFM images of the samples and modified Franck−Condon analysis, description of the TA spectrometer, formulas of optical densities, cross sections and excitation densities, raw TA data for comparison, energy transfer from amorphous P3HT to aggregated P3HT, quantum yields, P3HT cation absorption spectrum in solution, TA measurements of P3HT in dilute solutions, bimolecular recombination, anisotropy measurements, exemplary 2D gap-free TA maps. This material is available free of charge via the Internet at http://pubs.acs.org.

■ AUTHOR INFORMATION

Corresponding Author
riedle@physik.uni-muenchen.de

■ ACKNOWLEDGMENT

The authors thank H. Bässler (Universität Bayreuth, Germany) for very fruitful discussions, P. Rupp and M. Bradler for experimental support, R. Tautz for help with the P3HT oxidation in solution, M. Algasinger and T. Antesberger for synthesizing the poly-Si, H. Wiggers for providing the Si-ncs, and J. Gmeiner for the synthesis of RRa-P3HT. D.H. is grateful to Studienstiftung des Deutschen Volkes. S.N. is thankful to the Karl-Max von Bauernfeind-Verein and the International Graduate School "Material Science for Complex Interfaces (CompInt)" of the Technische Universität München, Germany. C.S. and A.K. are thankful to the Graduiertenkolleg 1640 of the DFG. E.R. acknowledges funding from the SFB 749.

■ REFERENCES

(1) Thompson, B. C.; Fréchet, J. M. J. *Angew. Chem., Int. Ed.* **2008**, *47*, 58−77.
(2) Brabec, C. J.; Sariciftci, N. S.; Hummelen, J. C. *Adv. Funct. Mater.* **2001**, *11*, 15−26.
(3) Park, S. H.; Roy, A.; Beaupré, S.; Cho, S.; Coates, N.; Moon, J. S.; Moses, D.; Leclerc, M.; Lee, K.; Heeger, A. J. *Nat. Photonics* **2009**, *3*, 297−303.
(4) Gur, I.; Fromer, N. A.; Geier, M. L.; Alivisatos, A. P. *Science* **2005**, *3*, 462−465.
(5) Jabbour, G. E.; Doderer, D. *Nat. Photonics* **2010**, *4*, 604.
(6) Huynh, W. U.; Dittmer, J. J.; Alivisatos, A. P. *Science* **2002**, *295*, 2425−2427.
(7) Oosterhout, S. D.; Wienk, M. M.; van Bavel, S. S.; Thiedmann, R.; Koster, L. J. A.; Gilot, J.; Loos, J.; Schmidt, V.; Janssen, R. A. J. *Nat. Mater.* **2009**, *8*, 810−824.
(8) Liu, C.-Y.; Holman, Z. C.; Kortshagen, U. R. *Nano Lett.* **2009**, *9*, 449−452.
(9) Dietmueller, R.; Stegner, A. R.; Lechner, R.; Niesar, S.; Pereira, R. N.; Brandt, M. S.; Ebbers, A.; Trocha, M.; Wiggers, H.; Stutzmann, M. *Appl. Phys. Lett.* **2009**, *94*, 113301.
(10) Niesar, S.; Dietmueller, R.; Nesswetter, H.; Wiggers, H.; Stutzmann, M. *Phys. Status Solidi A* **2009**, *206*, 2775−2781.
(11) Beek, W. J. E.; Wienk, M. M.; Janssen, R. A. J. *Adv. Funct. Mater.* **2006**, *16*, 1112−1116.
(12) Briseno, A. L.; Holcombe, T. W.; Boukai, A. I.; Garnett, E. C.; Shelton, S. W.; Fréchet, J. M. C.; Yang, P. *Nano Lett.* **2010**, *10*, 334−340.
(13) Liu, J.; Kadnikova, E. N.; Liu, Y.; McGehee, M. D.; Fréchet, J. M. J. *J. Am. Chem. Soc.* **2004**, *126*, 9486−9487.
(14) Liu, J.; Tanaka, T.; Sivula, K.; Alivisatos, A. P.; Fréchet, J. M. J. *J. Am. Chem. Soc.* **2004**, *126*, 6550−6551.
(15) McDonald, S. A.; Konstantatos, G.; Zhang, S.; Cyr, P. W.; Klem, E. J. D.; Levina, L.; Sargent, E. H. *Nat. Mater.* **2005**, *4*, 138−142.
(16) Howard, I. A.; Mauer, R.; Meister, M.; Laquai, F. *J. Am. Chem. Soc.* **2010**, *132*, 14866−14876.
(17) Kim, Y.; Cook, S.; Tuladhar, S. M.; Choulis, S. A.; Nelson, J.; Durrant, J. R.; Bradley, D. D. C.; Giles, M.; McCulloch, I.; Ha, C.-S.; Ree, M. *Nat. Mater.* **2006**, *5*, 197−203.
(18) Peet, J.; Kim, J. Y.; Coates, N. E.; Ma, W. L.; Moses, D.; Heeger, A. J.; Bazan, G. C. *Nat. Mater.* **2007**, *6*, 497−500.
(19) Clarke, T.; Ballantyne, A. M.; Nelson, J.; Bradley, D. D. C.; Durrant, J. R. *Adv. Funct. Mater.* **2008**, *18*, 4029−4035.
(20) Spano, F. C. *J. Chem. Phys.* **2005**, *122*, 234701.
(21) Clark, J.; Silva, C.; Friend, R. H.; Spano, F. C. *Phys. Rev. Lett.* **2007**, *98*, 206406.
(22) McCullough, R. D.; Tristram-Nagle, S.; Williams, S. P.; Lowe, R. D.; Jayaraman, M. *J. Am. Chem. Soc.* **1993**, *115*, 4910−4911.
(23) Chen, T.-A.; Wu, X; Rieke, R. D. *J. Am. Chem. Soc.* **1995**, *117*, 233−244.
(24) Sirringhaus, H.; Brown, P. J.; Friend, R. H.; Nielsen, M. M.; Bechgaard, K.; Langeveld-Voss, B. M. W.; Spiering, A. J. H.; Janssen, R. A. J.; Meijer, E. W.; Herwig, P.; de Leeuw, D. M. *Nature* **1999**, *401*, 685−688.
(25) Österbacka, R.; An, C. P.; Jiang, X. M.; Vardeny, Z. V. *Science* **2000**, *287*, 839−842.

(26) Chang, J.-F.; Clark, J.; Zhao, N.; Sirringhaus, H.; Breiby, D. W.; Andreasen, J. W.; Nielsen, M. M.; Giles, M.; Heeney, M.; McCulloch, I. *Phys. Rev. B* **2006**, *74*, 115318.
(27) Mauer, R.; Kastler, M.; Laquai, F. *Adv. Funct. Mater.* **2010**, *20*, 2085–2092.
(28) Brabec, C. J.; Zerza, G.; Cerullo, G.; De Silvestri, S.; Luzzati, S.; Hummelen, J. C.; Sariciftci, S. *Chem. Phys. Lett.* **2001**, *340*, 232–236.
(29) Piris, J.; Dykstra, T. E.; Bakulin, A. A.; van Loosdrecht, P. H. M.; Knulst, W.; Trinh, M. T.; Schins, J. M.; Siebbeles, L. D. A. *J. Phys. Chem. C* **2009**, *113*, 14500–14506.
(30) Guo, J.; Ohkita, H.; Benten, H.; Ito, S. *J. Am. Chem. Soc.* **2010**, *132*, 6154–6164.
(31) Banerji, N.; Cowan, S.; Leclerc, M.; Vauthey, E.; Heeger, A. J. *J. Am. Chem. Soc.* **2010**, *132*, 17459–17470.
(32) Brabec, C. J.; Gowrisankar, S.; Halls, J. J. M.; Laird, D.; Jia, S.; Williams, S. P. *Adv. Mater.* **2010**, *22*, 3839–3856.
(33) Guo, J.; Ohkita, H.; Benten, H.; Ito, S. *J. Am. Chem. Soc.* **2009**, *131*, 16869–16880.
(34) Etzold, F.; Howard, I. A.; Mauer, R.; Meister, M.; Kim, T.-D.; Lee, K.-S.; Baek, N. S.; Laquai, F. *J. Am. Chem. Soc.* **2011**, *133*, 9469–9479.
(35) Grancini, G.; Polli, D.; Fazzi, D.; Cabanillas-Gonzalez, J.; Cerullo, G.; Lanzani, G. *J. Phys. Chem. Lett.* **2011**, *2*, 1099.
(36) Yoshino, K.; Hayashi, S.; Sugimoto, R. *Jpn. J. Appl. Phys.* **1984**, *23*, L899.
(37) Khan, A. L. T.; Sreearunothai, P.; Herz, L. M.; Banach, M. J.; Köhler, A. *Phys. Rev. B* **2004**, *69*, 085201.
(38) Campbell, A. R.; Hodgkiss, J. M.; Westenhoff, S.; Howard, I. A.; Marsh, R. A.; McNeill, C. R.; Friend, R. H.; Greenham, N. C. *Nano Lett.* **2008**, *8*, 3942–3947.
(39) Scharsich, C; Lohwasser, R.; Asawapirom, U.; Scherf, U.; Thelakkat, M.; Köhler, A. Manuscript submitted, 2011.
(40) Knipping, J.; Wiggers, H.; Rellinghaus, B.; Roth, P.; Konjhodzic, D.; Meier, C. *J. Nanosci. Nanotechnol.* **2004**, *4*, 1039–1044.
(41) Germack, D. S.; Chan, C. K.; Kline, R. J.; Fischer, D. A.; Gundlach, D. J.; Toney, M. F.; Richter, L. J.; DeLongchamp, D. M. *Macromolecules* **2010**, *43*, 3828–3836.
(42) Scholz, M.; Gjukic, M.; Stutzmann, M. *Appl. Phys. Lett.* **2009**, *94*, 012108.
(43) Megerle, U.; Pugliesi, I.; Schriever, C.; Sailer, C. F.; Riedle, E. *Appl. Phys. B: Lasers Opt.* **2009**, *96*, 215–231.
(44) Wilhelm, T.; Piel, J.; Riedle, E. *Opt. Lett.* **1997**, *22*, 1494–1496.
(45) Riedle, E.; Beutter, M.; Lochbrunner, S.; Piel, J.; Schenkl, S.; Spörlein, S.; Zinth, W. *Appl. Phys. B: Lasers Opt.* **2000**, *71*, 457–465.
(46) Bradler, M.; Baum, P.; Riedle, E. *Appl. Phys. B: Lasers Opt.* **2009**, *97*, 561–574.
(47) Clark, J.; Chang, J.-F.; Spano, F. C.; Friend, R. H.; Silva, C. *Appl. Phys. Lett.* **2009**, *94*, 163306.
(48) Xu, B.; Holdcroft, S. *Macromolecules* **1993**, *26*, 4457–4460.
(49) Singh, R. K; Kumar, J.; Singh, R.; Kant, R.; Rastogi, R. C.; Chand, S.; Kumar, V. *New J. Phys.* **2006**, *8*, 112.
(50) Schueppel, R.; Schmidt, K.; Uhrich, C.; Schulze, K.; Wynands, D.; Brédas, J. L.; Brier, E.; Reinold, E.; Bu, H.-B.; Baeuerle, P.; Maennig, B.; Pfeiffer, M.; Leo, K. *Phys. Rev. B* **2008**, *77*, 085311.
(51) Jian, X. N.; Österbacka, R.; Korovyanko, O.; An, C. P.; Horovitz, B.; Janssen, R. A. J.; Vardeny, Z. V. *Adv. Funct. Mater.* **2002**, *12*, 587–597.
(52) van Hal, P. A.; Christiaans, M. P. T.; Wienk, M. M.; Kroon, J. M.; Janssen, R. A. J. *J. Phys. Chem. B* **1999**, *103*, 4352.
(53) Albert-Seifried, S; Friend, R. H. *Appl. Phys. Lett.* **2011**, *98*, 223304.
(54) Klessinger, M.; Michl, J. *Excited States and Photochemistry of Organic Molecules*; VCH Publishers, Inc.: New York, 1995; p 36.
(55) Köhler, A.; Bässler, H. *Mater. Sci. Eng., R* **2009**, *66*, 71–109.
(56) Ohkita, H.; Cook, S.; Astuti, Y.; Duffy, W.; Tierny, S.; Zhang, W.; Heeney, M.; McCulloch, I.; Nelson, J.; Bradley, D. D. C.; Durrant, J. R. *J. Am. Chem. Soc.* **2008**, *130*, 3030–3042.
(57) Tsoi, W. C.; Spencer, S. J.; Yang, L.; Ballantyne, A. M.; Nicholson, P. G.; Turnball, A.; Shard, A. G.; Murphy, C. E.; Bradley, D. D. C.; Nelson, J.; Kim, J.-S. *Macromolecules* **2011**, *44*, 2944–2952.

(58) Niesar, S.; Stegner, A. R.; Pereira, R. N.; Hoeb, M.; Wiggers, H.; Brandt, M. S.; Stutzmann, M. *Appl. Phys. Lett.* **2010**, *96*, 193112.
(59) Houston, P. L. *Chemical Kinetics and Reaction Dynamics*; Dover Publications, Inc.: Mineola, NY, 2001; pp 56–58.
(60) Middleton, C. T.; de La Harpe, K.; Su, C.; Law, Y. K.; Crespo-Hernández, C. E.; Kohler, B. *Annu. Rev. Phys. Chem.* **2009**, *60*, 217.
(61) Polli, D.; Altoè, P.; Weingart, O.; Spillane, K. M.; Manzoni, C.; Brida, D.; Tomasello, G.; Orlandi, G.; Kukura, P.; Mathies, R. A.; Garavelli, M.; Cerullo, G. *Nature* **2010**, *467*, 440.
(62) Chudoba, C.; Lutgen, S.; Jentzsch, T.; Riedle, E.; Woerner, M.; Elsaesser, T. *Chem. Phys. Lett.* **1995**, *240*, 35.
(63) Sobolewski, A. L.; Domcke, W.; Hättig, C. *J. Phys. Chem. A* **2006**, *110*, 6301.
(64) Petersson, J.; Eklund, M.; Davidsson, J.; Hammarström, L. *J. Am. Chem. Soc.* **2009**, *131*, 7940.
(65) Domcke, W.; Yarkony, D. R.; Köppel, H. *Conical Intersections: Electronic Structure, Dynamics & Spectroscopy* (*Advanced Series in Physical Chemistry*); World Scientific Publishing Co., Inc.: Singapore, 2004.
(66) Köppel, H.; Domcke, W.; Yarkony, D. R. *Conical Intersections: Theory, Computation and Experiments* (*Advanced Series in Physical Chemistry*); World Scientific Publishing Co., Inc.: Singapore, 2011.
(67) Ruseckas, A.; Wood, P.; Samual, I. D. W.; Webster, G. R.; Mitchell, W. J.; Burn, P. L.; Sundström, V. *Phys. Rev. B* **2005**, *72*, 115214.
(68) Hwang, I.; Scholes, G. D. *Chem. Mater.* **2011**, *23*, 610–620.
(69) Lin, H.; Tian, Y.; Zapadka, K.; Persson, G.; Thomsson, D.; Mirzov, O.; Larsson, P.-O.; Widengren, J.; Scheblykin, I. G. *Nano Lett.* **2009**, *9*, 4456.
(70) Laquai, F.; Park, Y.-S.; Kim, J.-J.; Basché, T. *Macromol. Rapid Commun.* **2009**, *30*, 1203–1231.
(71) Movaghar, B.; Grünewald, M.; Ries, B.; Bässler, H.; Würtz, D. *Phys. Rev. B* **1986**, *33*, 5545–5554.
(72) Gulbinas, V.; Zaushitsyn, Y.; Sundström, V.; Hertel, D.; Bässler, H.; Yartsev, A. *Phys. Rev. Lett.* **2002**, *89*, 107401.
(73) Shuttle, C. G.; O'Regan, B.; Ballantyne, A. M.; Nelson, J.; Bradley, D. D. C.; Durrant, J. R. *Phys. Rev. B* **2008**, *78*, 113201.
(74) Dogariu, A.; Vacar, D.; Heeger, A. J. *Phys. Rev. B* **1998**, *58*, 10218–10224.
(75) Köhler, A.; dos Santos, D. A.; Beljonne, D.; Shuai, Z.; Brédas, J,-L.; Kraus, A.; Müllen, K.; Friend, R. H. *Nature* **1998**, *392*, 903–906.
(76) Ferguson, A. J.; Kopidakis, N.; Shaheen, S. E.; Rumbles, G. *J. Phys. Chem. C* **2008**, *112*, 9865–9871.
(77) Howard, I. A.; Hodgkiss, J. M.; Zhang, X.; Kirov, K. R.; Bronstein, H. A.; Williams, C. K.; Friend, R. H.; Westenhoff, S.; Greenham, N. C. *J. Am. Chem. Soc.* **2010**, *132*, 328–335.
(78) Shuttle, C. G.; Hamilton, R.; O'Regan, B. C.; Nelson, J.; Durrant, J. R. *Proc. Natl. Acad. Sci. U.S.A.* **2010**, *107*, 16448–16452.
(79) Mauer, R.; Howard, I. A.; Laquai, F. *J. Phys. Chem. Lett.* **2010**, *1*, 3500–3505.
(80) Frolov, S. V.; Bao, Z.; Wohlgenannt, M.; Vardeny, Z. V. *Phys. Rev. Lett.* **2000**, *85*, 2196–2199.
(81) Stevens, M. A.; Silca, C.; Russell, D. M.; Friend, R. H. *Phys. Rev. B* **2001**, *63*, 165213.
(82) Silva, C.; Dhoot, A. S.; Russell, D. M.; Stevens, M. A.; Arias, A. C.; MacKenzie, J. D.; Greenham, N. C.; Friend, R. H. *Phys. Rev. B* **2001**, *64*, 125211.
(83) Arkhipov, V. I.; Emelianova, E. V.; Bässler, H. *Phys. Rev. Lett.* **1999**, *82*, 1321–1324.
(84) Zenz, C.; Lanzani, G.; Cerullo, G.; Graupner, W.; Leising, G.; Scherf, U.; DeSilvestri, S. *Synth. Met.* **2001**, *116*, 27–30.
(85) Parkinson, P.; Müller, C.; Stingelin, N.; Johnson, M. B.; Herz, L. M. *J. Phys. Chem. Lett.* **2010**, *1*, 2788–2792.
(86) Westenhoff, S.; Beenken, W. J. D.; Friend, R. H.; Greenham, N. C.; Yartsev, A.; Sundström, V. *Phys. Rev. Lett.* **2006**, *97*, 166804.
(87) Hodgkiss, J. M.; Campbell, A. R.; Marsh, R. A.; Rao, A.; Albert-Seifried, S.; Friend, R. H. *Phys. Rev. Lett.* **2010**, *104*, 177701.
(88) Kim, J. Y.; Lee, K.; Coates, N. E.; Moses, D.; Nguyen, T.-Q.; Dante, M.; Heeger, A. J. *Science* **2007**, *317*, 222–225.
(89) Gilot, J.; Wienk, M. M.; Janssen, R. A. J. *Appl. Phys. Lett.* **2007**, *90*, 143512.

Role of Structural Order and Excess Energy on Ultrafast Free Charge Generation in Hybrid Polythiophene/Si Photovoltaics Probed in Real Time by Near-Infrared Broadband Transient Absorption

Daniel Herrmann[1], Sabrina Niesar[2], Christina Scharsich[3], Anna Köhler[3], Martin Stutzmann[2], and Eberhard Riedle[1]

[1] Lehrstuhl für BioMolekulare Optik, Ludwig-Maximilians-Universität München, Oettingenstr. 67, 80538 München, Germany

[2] Walter Schottky Institut, Technische Universität München, Am Coulombwall 4, 85748 Garching, Germany

[3] Lehrstuhl EP II, Universität Bayreuth, Universitätsstr. 30, 95440 Bayreuth, Germany

----
Supporting Information

Contact Information:

Prof. Eberhard Riedle, LMU München
Phone: +49(0)89/2180-9210
Email: riedle@physik.uni-muenchen.de

Prof. Martin Stutzmann, TU München
Phone: +49(0)89/289-12760
Email: Stutz@wsi.tum.de

Prof. Anna Köhler, University of Bayreuth
Phone: +49(0)921/55-2600
Email: anna.koehler@uni-bayreuth.de

Basic Optical and Structural Characterization of P3HT

The basic optical characterization of the thin film samples was done by UV-Vis-NIR absorbance measurements recorded with a Perkin Elmer Lambda 750 spectrophotometer. To differentiate between absorption and scatter, some spectra were recorded with a Perkin Elmer Lambda 900 equipped with an integrating sphere. Visible photoluminescence spectra were recorded with a Spex Fluorolog 2 fluorimeter. To gain knowledge on the morphology of the samples atomic force microscopy (AFM) images were taken with a Veeco MultiMode Nanoscope V microscope in tapping mode (Figs. 2,S1). The scan size was 10 µm x 10 µm.

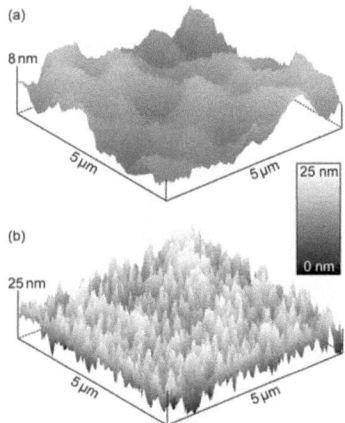

Figure S1. Corresponding topographical 3D AFM images of spin-coated films of (a) RRa-P3HT and (b) aggRR-P3HT on fused silica.

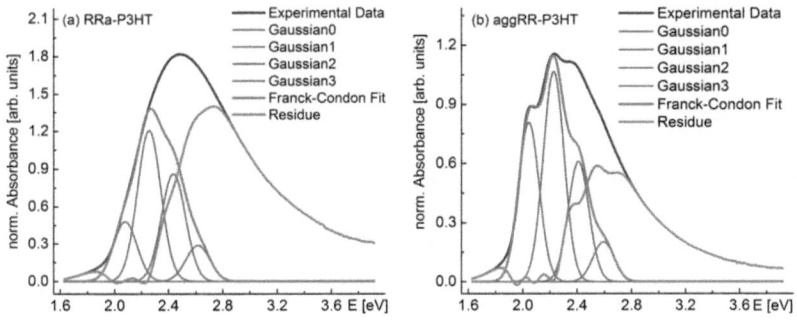

Figure S2. Modified Franck-Condon fit to the absorption spectra of (a) RRa-P3HT and (b) aggRR-P3HT taken from Fig. 4(a).

With the assumption that the absorption coefficient of aggregated chains is 1.39 times that of coiled chains [S1,S2], a percentage of (38 ± 5)% aggregates for the aggRR-P3HT sample and of (24 ± 5)% aggregates for the RRa-P3HT sample can be derived with a modified Franck-Condon fit to the absorbance spectra as indicated in Fig. S2.

Description of the TA Spectrometer

To investigate the nature of the photoexcitations and their inherent kinetics after visible excitation we used a novel pump-probe spectrometer that allows for ultrafast broadband (UV-Vis-NIR) transient absorption (TA) experiments at room temperature with a time resolution of 40 fs. The UV-Vis part of the setup employing a Ti:Sa chirped-pulse amplifier system (CPA 2001, ClarkMXR) with 1 kHz repetition rate as common fundamental light source for pump and probe generation has been described previously. It was extended to allow the investigation of thin film samples and to expand the detection range into the NIR (Fig. S3). A single-stage noncollinear optical parametric amplifier (NOPA) was used as pump source providing ultrashort pump pulses with tunable wavelength. The NOPA allows a very broad tuning of the pump wavelength providing the possibility of a selective excitation from 450 to 760 nm. For the P3HT-based samples investigated in this work, excitation wavelengths of 450, 518, 535, 555, 600 and 720 nm were selected.

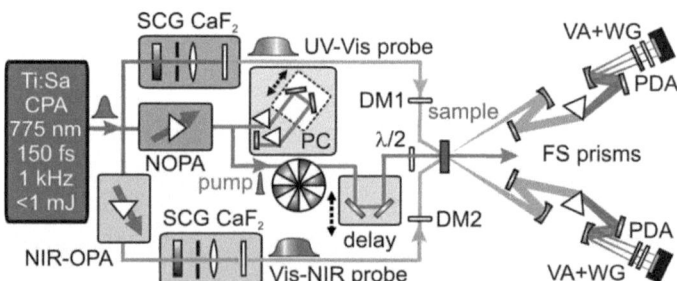

Figure S3. Experimental layout of the novel ultrafast UV-Vis-NIR TA pump-probe spectrometer. NOPA: noncollinear optical parametric amplifier, SCG: supercontinuum generation, PC: prism compressor, DM: dichroic mirror, VA: variable attenuator, WG: wire-grid polarizer, PDA: photodiode array.

The NOPA pulses were compressed with a fused silica prism compressor (PC) and showed a Fourier-limited 15 fs FWHM pulse duration, measured at the sample position with a second-harmonic generation scanning autocorrelator. The pump energy was adjusted to 4-60 nJ using a combination of an achromatic $\lambda/2$ waveplate and broadband wire-grid polarizer (Moxtek Inc.). The energy stability was typically 1.5% RMS. A chopper wheel in the pump beam blocked every second excitation pulse such that changes in the optical density (ΔOD) of the sample could be recorded. The pump beam was led through a 2 ns delay stage. A $\lambda/2$ waveplate in front of the sample was used to adjust the linear pump polarization relative to the linear probe polarization.

The UV-Vis and a Vis-NIR broadband probe source were employed in subsequent measurements. For gap-free TA measurements up to the NIR spectral range, a novel probe setup was developed employing a combination of two supercontinuum generation (SCG) stages and OPA (Fig. S3). First in a NIR-OPA, approximately 1 µJ light from the Ti:Sa amplifier was focused into a 4 mm YAG crystal to generate a supercontinuum with high NIR content [46]. The SC was subsequently amplified at 1180 nm in a collinear OPA stage employing a 3 mm long type-I BBO crystal (phase-matching angle: 20°) and pumped by the fundamental light from the Ti:Sa. Second, about 1-2 µJ of the amplified signal at 1180 nm were focused into a 5 mm long circularly translated CaF_2 crystal for SCG. After passing a short pass dichroic mirror (DM) deposited on a 1 mm fused silica substrate to block the intense light around 1180 nm and an OD filter, the broadband probe ranged from 415 to 1150 nm limited by the transmission of the DM.

The pump and probe were weakly focused into the sample leading to a 210 µm and 110 µm $1/e^2$ beam diameter at the sample, respectively. The weak focusing leads to an ensemble averaging over the finely grained morphology of the thin films to mimic their usage as photovoltaic device and to ensure low local excitation densities. We routinely verified that the samples did not degrade due to the prolonged illumination by the pump pulses and only observed signal degradation when storing the samples over several months at ambient conditions in the lab.

A broadband multichannel detection of the TA signals employing a fused silica (FS) prism polychromator and a silicon photodiode array (PDA) based camera with 512 pixels is used. Typically, a probe noise below 2% and a probe detection sensitivity of $\Delta OD \sim 10^{-4}$ over the entire spectral range were achieved. Selected measurements were repeated on later days and with additional samples of the same nominal composition. No differences were found within experimental accuracy.

All TA measurements were performed from -10 ps up to 2 ns delay in small steps and additionally at fixed delays of 3, 4, and 6 ns. 3 runs with averaging over 1500 pump on/off couples per delay step were recorded and averaged. The pump beam was led through a carefully aligned 2 ns delay stage with a maximum pump pointing deviation at the sample of 2 μm over the whole delay range. The vertically aligned sample was not moved during the recordings and could be adjusted in all directions. The spectral resolution at the detector was better than 3 nm/pixel. To minimize pump straylight originating from the solid film samples, a broadband wire-grid polarizer (WG) was placed in front of the detector. This WG was aligned with maximum transmission for the probe polarization and the pump polarization was adjusted to be perpendicular to the probe polarization in this case.

For accurate calibration of the time zero, the raw data was corrected for the chirp of the probe pulses by adjusting the zero time at half rise of the signals. Additionally, we could show that the pump-probe setup is interferometrically stable and thus offers the alternative to employ a periodic modulation due to spectral interference between pump straylight and probe to extract the delay via Fourier-transformation. We have explicitly compared both options and found that the deviation of their result is negligible and enable calibration of the zero delay with a few-fs accuracy. Overall, the time resolution of the ultrabroadband TA measurements was 40 fs, in particular because we do not need to measure over the spectral region of the SCG fundamental where a time jump is present in the broadband probe pulse. This is considerably better than for measurements with solutions due to the effective lack of a group velocity mismatch contribution in the 100 nm thin films.

Formulas of Optical Densities, Cross-sections and Excitation Densities

The Beer-Lambert law is a linear relationship between absorbance change ΔOD (optical density, OD) and photoinduced concentration change Δc of the corresponding species generated upon photoexcitation (e.g. polarons) with its inherent molar extinction coefficient ε_M in L mol^{-1} cm^{-1}

$$\Delta OD(t) = \varepsilon_M \cdot \Delta c(t) \cdot l = -\log(T(t)/T_0) \tag{1}$$

where l is the sample thickness. The transient change in optical density ΔOD is linked to the transmission T after photoexcitation and T_0 without excitation. T and T_0 are recorded during the TA measurements.

The carrier concentration c in mol/L is connected to the carrier density n in cm^{-3} via

$$c = \frac{1000}{N_A} n, \quad (2)$$

where N_A is Avogadro's constant. Therefore the polaron molar extinction coefficient can be derived from the TA data

$$\varepsilon_M = \frac{\Delta OD_{300fs}}{[P3HT^+]_{300fs} \cdot l} \cdot \frac{N_A}{1000}, \quad (3)$$

where for the carrier density, the initial photoinduced charge density $[P3HT^+]_{300fs}$ at the maximum of the polaron signal (at 300 fs) is taken. For its calculation, the polaron cross-section is needed.

From the TA data we can estimate the polaron cross-section σ

$$\sigma = \frac{\Delta T}{T \cdot N}, \quad (4)$$

where N is the areal density of excited states. N is equal to the pump photon number per square centimeter multiplied by the fraction of photons absorbed, which is determined by the absorbance of the sample at the pump wavelength. Employing the polaron cross-section, we determine the initial charge density $[P3HT^+]_{300fs}$ at the maximum of the polaron signal

$$n(t=300\ fs) = [P3HT^+]_{300fs} = \frac{|\Delta T/T|_{300fs}}{\sigma \cdot l}, \quad (5)$$

where $|\Delta T/T|_{300fs}$ is the initial magnitude of the polaron TA signal at 300 fs.

The cross-section is related to the absorption coefficient α in m^{-1} and the extinction coefficient via

$$\sigma = \frac{\alpha}{n} = \frac{\varepsilon_M \cdot c}{\log(e) \cdot n} = \frac{\varepsilon_M \cdot 1000}{\log(e) \cdot N_A}. \quad (6)$$

Raw TA Data for Comparison

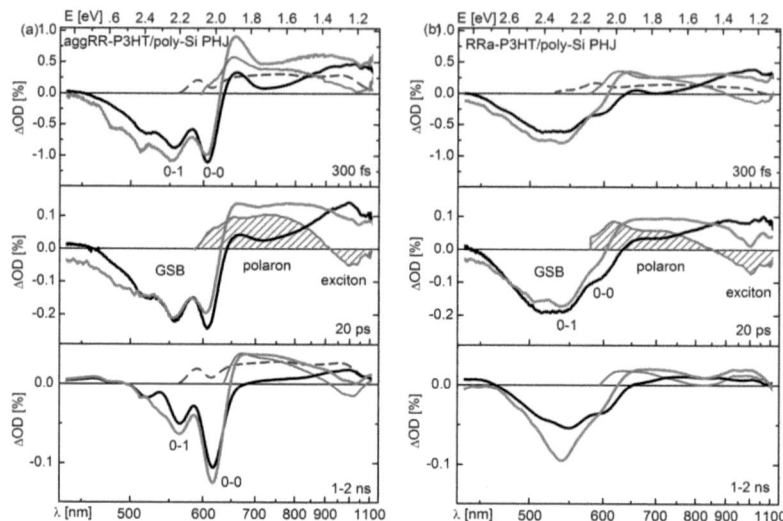

Figure S4. (a) Raw TA spectra of pure aggRR-P3HT (black) and aggRR-P3HT/poly-Si PHJ (red). (b) Raw TA spectra of pure RRa-P3HT (black) and RRa-P3HT/poly-Si PHJ (red). All at the nominal same excitation of 9 µJ/cm^2 at 518 nm. The difference spectra (green) are compared to the individual P3HT film polaron absorption spectrum (blue dashes).

Energy Transfer from amorphous P3HT to aggregated P3HT

When considering the vibrational structure displayed in Fig. 5 of the manuscript for different delay times, we observe that the ratio 0-0 to 0-1 decreases until 13 ps and subsequently increases again. In addition, the relative intensity of the GSB signal around 500 nm reduces with time. These changes imply that some dynamic process must take place in P3HT after photoexcitation. We recall that the P3HT absorption spectrum, and similarly the GSB spectrum, arises from a superposition of contributions of coils (around 450-500 nm) and aggregates (around 600 nm). In addition, within the aggregate absorption spectrum, a high 0-0 peak signal implies planar chains with long conjugation length and with weak excitonic coupling to their neighbors, while a lower 0-0 peak suggests chains with somewhat shorter conjugation length and concomitantly stronger excitonic coupling [S3]. Furthermore, the reduction of the relative GSB intensity around 500 nm with time suggests that the ground state of the coiled chains recovers faster than the ground state of the aggregated chains, for example by energy transfer from coiled to aggregated chains. To quantify on which timescale this energy transfer takes place, we have displayed the ratio between the GSB intensity at 480 nm for excitation at 450 nm to excitation at 600 nm (Fig. S5), using the data displayed in Fig. 8 of the manuscript. This ratio reaches a constant value after about 13-20 ps. Fig. S5 reveals a forward 1/e energy transfer time of about 3 ps. It is interesting to note that the same 1/e time of 3 ps had been found for the energy transfer from a glassy phase to a planarized phase in the blue-light-emitting poly(9,9-dioctylfluorene) [S4].

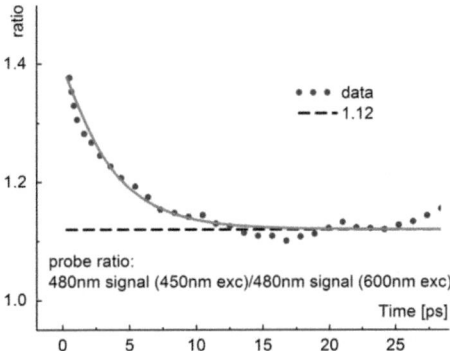

Figure S5. The ratio of 480 nm GSB signal for 450 nm to the 480 nm GSB signal for 600 nm excitation shows a 1/e energy transfer time from coils to aggregates of 3 ps derived from a monoexponential fit. The energy transfer is completed at around 13-20 ps.

In order to analyse the temporal changes in the 0-0 intensity shown in Fig. 6 of the manuscript in full quantitative detail, the contributions from coiled and aggregated chains would need to be deconvoluted. This is beyond the scope of this paper. However, the change from a relatively high 0-0 peak (300 fs) to a lower one (20 ps) and again to a high 0-0 peak (1-2 ns) is obvious. It suggests the excitation to be initially delocalized, then more localized and subsequently again more delocalized. Whether these changes are due to structural rearrangements of polymer chains such as torsional relaxation [S5], or whether they are due to spectral diffusion to more conjugated chain segments, or due to a combination of both effects[S6], cannot be distinguished on the basis of these data.

Quantum Yields

The PIA is determined by singlet-excitons and polarons in P3HT. The maximum quantum yields of charges can be derived from the TA spectra in Fig. 5, accounting for the observed photovoltaic conversion pathways shown in Scheme S1 [S7].

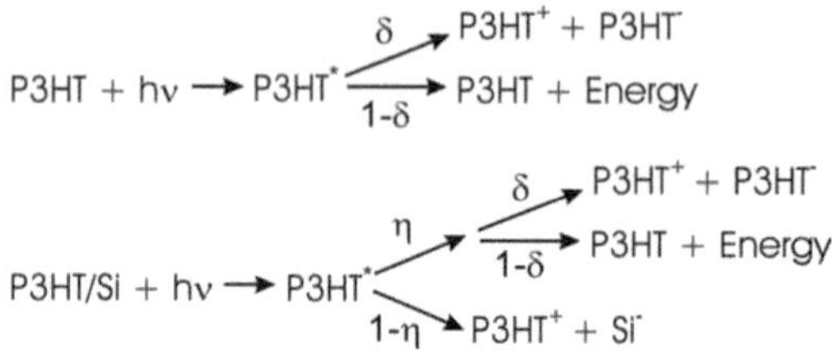

$$P3HT + h\nu \rightarrow P3HT^* \begin{array}{c} \xrightarrow{\delta} P3HT^+ + P3HT^- \\ \xrightarrow{1-\delta} P3HT + \text{Energy} \end{array}$$

$$P3HT/Si + h\nu \rightarrow P3HT^* \begin{array}{c} \xrightarrow{\eta} \begin{array}{c} \xrightarrow{\delta} P3HT^+ + P3HT^- \\ \xrightarrow{1-\delta} P3HT + \text{Energy} \end{array} \\ \xrightarrow{1-\eta} P3HT^+ + Si^- \end{array}$$

Scheme S1. Yields of transient species generated by photoexcitation of pure P3HT and P3HT/Si heterojunctions.

According to our findings, exciting pure P3HT with pump photons in the absorption range initially leads to the formation of singlet-excitons. These Frenkel-type excitons can dissociate into positive ($P3HT^+$) and negative ($P3HT^-$) P3HT polarons with quantum yield δ and the remaining photoexcitations can decay to the ground state with yield $1-\delta$ through release of vibrational energy (P3HT + Energy). In the P3HT/Si heterojunction, a proportion η of the photons is absorbed so far away from the hybrid heterojunction as to lead to the same result as in pure P3HT. The remaining fraction ($1-\eta$) of photons is absorbed near a P3HT/Si interface to yield ultrafast electron transfer to the Si with 140 fs forward transfer time. To determine the quantum yields (Tab. S1), we consider the TA signal magnitude at 300 fs due to excitons in aggRR-P3HT/poly-Si PHJ, which typically has about 0.75 times the magnitude as for pure aggRR-P3HT. Therefore, the fraction of photoexcitations that initially do not sense the Si component is taken as $\eta=0.75$. The maximal magnitude of the TA signal due to polarons at 300 fs is typically about 2.2 times that for pure P3HT, which results in $\delta = \dfrac{1-\eta}{2.2-\eta} = 0.17$.

Sample	$P3HT^+$	$P3HT^-$	P3HT+E	Si^-
aggP3HT	0.17	0.17	0.83	0
aggP3HT/Si	0.38	0.13	0.62	0.25

Table S1. Initial quantum yields of charges in pure P3HT and aggRR-P3HT/poly-Si PHJ

TA Measurements of RR-P3HT in Dilute Solutions

To identify the PIA of singlet-excitons, we performed TA measurements of RR-P3HT in dilute solutions of toluene, chloroform and 1,2-dichlorobenzene, where the intermolecular distance is too high for P3HT cation formation (Fig. S7). In this condition, the inherent lifetime of singlet-excitons can be studied. It is found to be significantly increased compared to the film, enabling their radiative decay (SE) and ISC to the triplet state on the ns timescale.

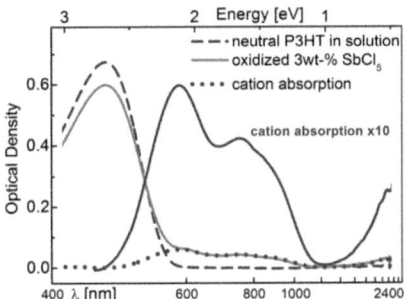

Figure S6. (a) Absorbance spectra of pure P3HT (blue dashes), chemically oxidized P3HT in 1,2-dichlorobenzene solution after adding 3wt-% pentachloroantimonate (SbCl$_5$) (solid grey curve) and P3HT cation absorption (brown).

P3HT in solution can be chemically oxidized employing a strong oxidant to obtain the P3HT cation signature. Fig. S6 shows the absorbance spectrum of chemically oxidized P3HT in 1,2-dichlorobenzene solution after adding 3wt-% pentachloroantimonate (SbCl$_5$). From this we obtained the P3HT cation (P3HT$^+$) absorption spectrum as follows. Adding SbCl$_5$ decreases the absorbance of the neutral P3HT molecule (blue dashes) and increases the cation absorbance (brown dots). Saturation occurs at adding 10wt-% SbCl$_5$. The difference between the blue dashed line scaled to the peak of the gray solid curve in Fig. S6 and the gray solid curve reveals the P3HT cation absorption (brown dots and brown solid line) ranging from 470 to about 1200 nm and increasing again up to the mid-infrared spectral region. The absorption spectrum of RR-P3HT in dilute solution (Fig. S6) shows reduced conjugation and hindered aggregation compared to RR-P3HT film (Fig. 3(b))[S3].

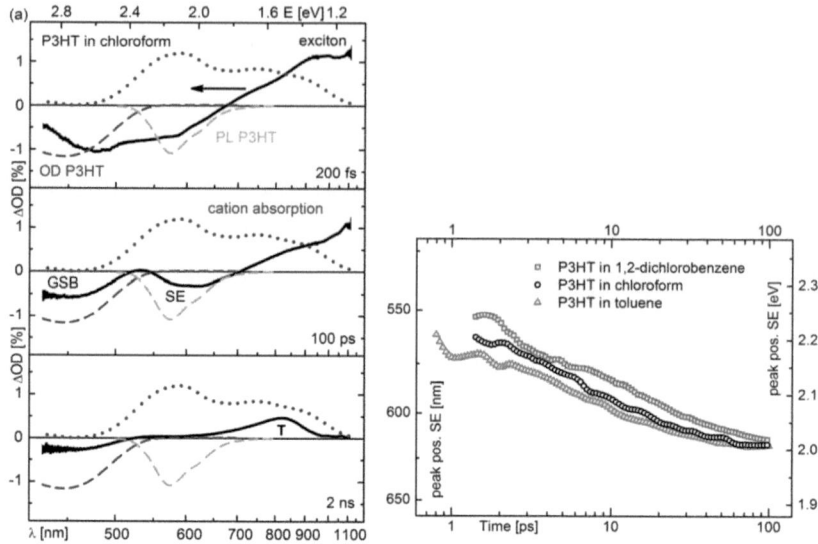

Figure S7. (a) TA spectra (black solid curves) of RR-P3HT dissolved in chloroform at 200 fs, 100 ps and 2 ns. The cation absorption spectrum (brown dots) as well as the P3HT absorption (blue dashed curve) and PL spectra (orange dashed curve) are scaled to the transient absorption spectrum to identify GSB, SE and singlet-/triplet-exciton absorption. Excitation: 15 µJ/cm² at 475 nm. (b) Time evolution of the SE center spectral position for P3HT in 1,2-dichlorobenzene (green squares), chloroform (black circles) and toluene (red triangles).

Fig. S7(a) shows TA spectra (black solid curves) of RR-P3HT in dilute solutions of chloroform at 200 fs, 100 ps and 2 ns with low excitation fluence of 15 µJ/cm² at 475 nm pump wavelength. In order to cancel out effects of molecule orientation in solution on the dynamics, the polarization direction of the linearly polarized pump pulse was adjusted at the magic angle of 54.7° with respect to that of the probe pulse. For comparison, the P3HT absorption (blue solid curve) and PL (orange solid curve) spectra as well as the P3HT cation absorption spectrum are incorporated in Fig. S7(a). The TA spectra reveal signatures of GSB and SE ($\Delta OD < 0$) as well as signatures of PIA ($\Delta OD > 0$) at the expected spectral positions. The GSB shows a more rapid decay at longer probe wavelengths (lower energy), which we attribute to a faster decay of excited longer conjugations that are less stabilized by interchain interactions [S3]. Fig. S7 shows that the SE spectrally shifts with time towards lower probe photon energy (longer probe wavelength) by about 0.22 eV until it vanishes at around 1 ns, whereas the total shift depends on the solvent polarity. We note that only broadband probe detection of the TA signals allows for observing this relaxation process which other-wise would be misinterpreted as a decay process in case of single-wavelength probe detection. We assign this effect to funneling of intrachain excitons to lower-energy sites [S5], torsional relaxation of the polymer backbone in the excited state [S5,S6] and solvation effects [S8,S9].

It should be noted, that no signatures of P3HT cation absorption are present in Fig. S7(a). The early PIA signature above 750 nm (1.65 eV) in Fig. S7 is assigned to singlet-exciton absorption in agreement with the results of Fig. 5, taking into account the energetic shift of the signals between film and solution. At around 2 ns, a distinct PIA peak at 1.49 eV (830 nm) remains, which stays constant at least up to 6 ns (the maximum investigated delay). Consequently, this signature can be assigned to a PIA of a P3HT triplet state, which confirms and adds to recent theoretical predictions [S10]. We attribute this observation to intersystem crossing (ISC) with a ~1 ns-1 rate from the singlet-exciton to the triplet state [S11,S12].

Bimolecular Recombination

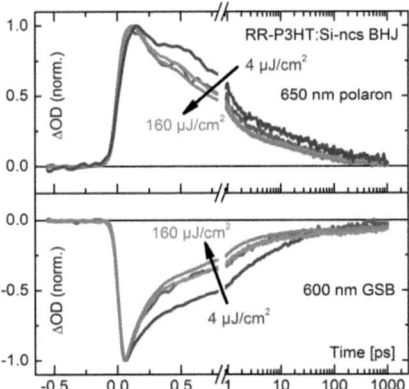

Figure S8. (a) Decay of TA signals for increased excitation fluences (4, 30, 60 and 160 µJ/cm² at 480 nm) reveals bimolecular nongeminate recombination in case of RR-P3HT:Si-ncs BHJ.

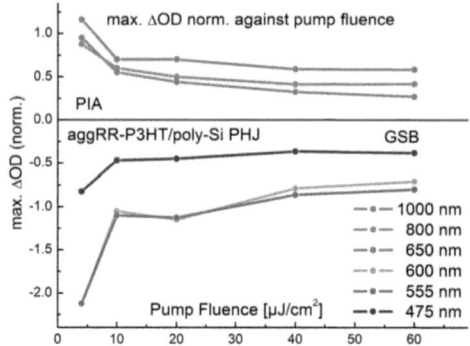

Figure S9. Transient signal magnitudes normalized with respect to excitation fluence plotted against the excitation fluence for aggRR-P3HT/Si PHJ excited at 518 nm. The polarons (650, 800 nm) show the same intensity dependence as the excitons (1000 nm) and are thus generated from excitons.

Time-dependent Anisotropy

The TA spectra in Fig. 5 were recorded with perpendicular (pp) pump and probe polarization to allow for optimum straylight suppression. To study the time-dependent anisotropy r(t) of polarons and singlet-excitons via Eq. 15, additional traces were recorded with both polarizations parallel (pa) to each other.

$$r(t) = \frac{\Delta OD_{pa} - \Delta OD_{pp}}{\Delta OD_{pa} + 2 \cdot \Delta OD_{pp}} = 0.2 \cdot (3 \cdot \cos^2 \theta - 1) \tag{15}$$

From the anisotropy, the the transition dipole orientation angle θ can be derived. The excitation fluence was kept low at 9 µJ/cm². Fig. S10(a) compares the anisotropy spectra of pure aggRR-P3HT (green solid curve) and aggRR-P3HT/poly-Si PHJ (blue solid curve) with the corresponding TA spectrum of pure aggRR-P3HT (black dashes) all at 300 fs with 518 nm excitation. Fig. S10(b) shows the time-dependent anisotropy decay of the transient species.

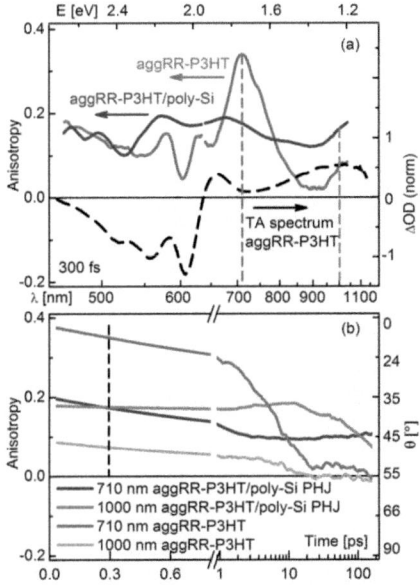

Figure S10. (a) Anisotropy spectrum at 300 fs of pure aggRR-P3HT (green solid curve) and aggRR-P3HT/poly-Si PHJ (blue solid curve) compared to the corresponding normalized TA spectrum (black dashes) of pure aggRR-P3HT (Fig. 6(a)). (b) Transient anisotropy decay and corresponding transition dipole reorientation angle θ of polaron (710 nm) and singlet-exciton (1000 nm).

In case of pure aggRR-P3HT, the anisotropy at 710 nm probe wavelength within the characteristic polaron band is r = 0.39 immediately after photoexcitation, which corresponds to parallel pump and probe transition dipole moments. Subsequently, the anisotropy decays to r = 0.20 at 4 ps and levels off at r ~ 0.02 after 20 ps (Fig. S10(b)). The anisotropy at the characteristic exciton band (1000 nm) is r = 0.10 immediately after photoexcitation (Fig. S10(b)), subsequently decays to r = 0.05 at 1 ps and finally completely vanishes at 20 ps, which might indicate a random relative orientation of pump and probe transition dipole moments. The anisotropy spectrum of pure aggRR-P3HT at 300 fs reveals a distinct peak from 620 to 910 nm, which matches the spectral range of polaron absorption in Fig. 5(a). We therefore assign the distinct peak around 710 nm with high anisotropy to polarons in pure aggRR-P3HT that are localized on a chain and completely loose their initial polarization memory within 20 ps through energy migration [S13]. We attribute this anisotropy decay to hopping of charge carriers with a characteristic 1/e hopping time of 7 ps.

On the other hand, the anisotropy spectrum of the aggRR-P3HT/poly-Si PHJ at 300 fs reveals no distinct feature. The anisotropy at the characteristic polaron band (710 nm) in case of the aggRR-P3HT/poly-Si PHJ is $r = 0.19$ immediately after photoexcitation. Subsequently, the anisotropy decays to $r = 0.13$ at 1 ps and $r = 0.10$ at 20 ps, where it saturates (Fig. S10(b)). Fig. 7(a) reveals that the ultrafast charge transfer between aggRR-P3HT and Si leads to generation of mobile charge carriers. We therefore conclude that the ultrafast loss of anisotropy for polarons in the aggRR-P3HT/poly-Si PHJ compared to aggRR-P3HT is attributable to this ultrafast charge transfer leading to a rapid loss of polarization memory and uncorrelated charges. It should be noted that an excess electron in the inorganic semiconductor Si possesses a highly delocalized wavefunction.

Exemplary 2D Gap-free TA Maps

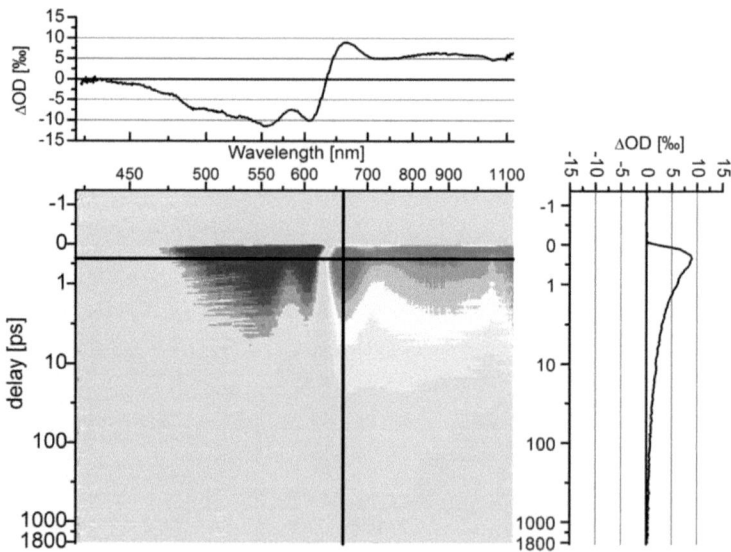

Figure S11(a). 2D map and lineouts for the TA measurement of aggRR-P3HT/Si PHJ excited at 518 nm with 9 µJ/cm².

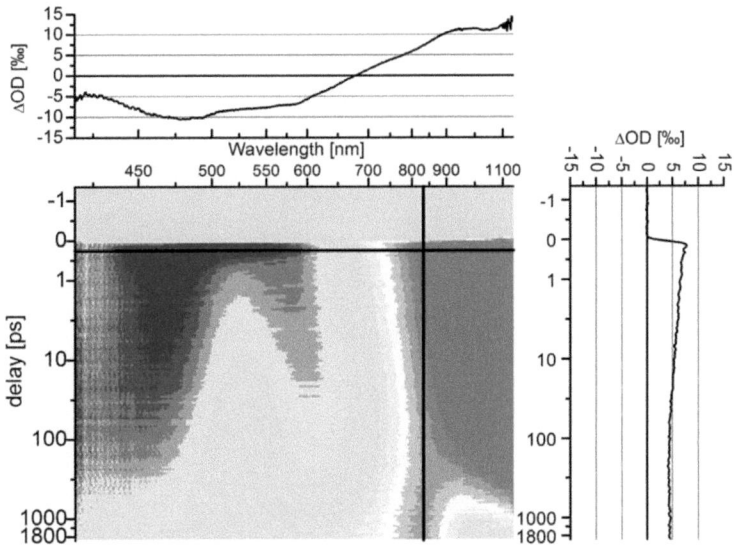

Figure S11(b). 2D map and lineouts for the TA measurement in Fig. S7(a).

(S1) Clark, J.; Chang, J.-F; Spano, F. C.; Friend, R. H.; Silva, C. Appl. Phys. Lett. **2009**, 94, 163306.

(S2) Scharsich, C; Lohwasser, R.; Asawapirom U.; Scherf U.; Thelakkat, M.; Köhler, A. (2011, submitted).

(S3) Clark, J.; Silva, C.; Friend, R. H.; Spano, F. C Phys. Rev. Lett. **2007**, 98, 206406.

(S4) Khan, A. L. T.; Sreearunothai, P.; Herz, L. M.; Banach, M. J.; Köhler, A. Phys. Rev. B **2004**, 69, 085201.

(S5) Parkinson, P.; Müller, C.; Stingelin, N,; Johnson, M. B.; Herz, L. M. J. Phys. Chem. Lett. **2010**, 1, 2788-2792.

(S6) Westenhoff, S.; Beenken, W. J. D.; Friend, R. H., Greenham, N. C.; Yartsev, A.; Sundström, V. Phys. Rev. Lett. **2006**, 97, 166804.

(S7) Piris, J.; Dykstra, T. E.; Bakulin, A. A.; van Loosdrecht, P. H. M.; Knulst, W.; Trinh, M. T.; Schins, J. M.; Siebbeles, L. D. A. J. Phys. Chem. C **2009**, 113, 14500-14506.

(S8) Horng, M. L.; Gardecki, J. A.; Papazyan, A.; Maroncelli, M. J. Phys. Chem. **1995**, 99, 17311-17337.

(S9) Reynolds, L.; Gardecki, J. A.; Frankland, S. J. V.; Horng, M. L.; Maroncelli, M. J. Phys. Chem. **1996**, 100, 10337-10354.

(S10) Köhler, A.; Bässler, H. Mater. Sci. Engineering R **2009**, 66, 71–109.

(S11) Kraabel, B.; Moses, D.; Heeger, A. J. J. Chem. Phys. **1995**, 103, 5102-5108.

(S12) Grebner, D.; Helbig, M.; Rentsch, S. J. Phys. Chem. **1995**, 99, 16991-16998.

(S13) Guo, J.; Ohkita, H.; Benten, H.; Ito, S. J. Am. Chem. Soc. **2009**, 131, 16869-16880.

Appendix B8

Few-Cycle Laser-Driven Electron Acceleration

K. Schmid, L. Veisz, F. Tavella, S. Benavides, R. Tautz, D. Herrmann, A. Buck, B. Hidding,
A. Marcinkevicius, U. Schramm, M. Geissler, J. Meyer-ter-Vehn, D. Habs, and F. Krausz

Reprinted with permission from
Physical Review Letters 102, 124801 (2009).

DOI: 10.1103/PhysRevLett.102.124801

http://link.aps.org/doi/10.1103/PhysRevLett.102.124801

Copyright © 2011, American Physical Society.

Few-Cycle Laser-Driven Electron Acceleration

K. Schmid,[1,2,*] L. Veisz,[1,*] F. Tavella,[1,3] S. Benavides,[1] R. Tautz,[1] D. Herrmann,[1] A. Buck,[1] B. Hidding,[4] A. Marcinkevicius,[1,5] U. Schramm,[6] M. Geissler,[7] J. Meyer-ter-Vehn,[1] D. Habs,[2] and F. Krausz[1]

[1]*Max-Planck-Institut für Quantenoptik, Hans-Kopfermann-Strasse 1, 85748 Garching, Germany*
[2]*Ludwig-Maximilians-Universität München, Am Coulombwall 1, D-85748 Garching, Germany*
[3]*Deutsches Elektronensynchrotron DESY/HASYLAB, Notkestrasse 85, 22607 Hamburg, Germany*
[4]*Heinrich-Heine-Universität Düsseldorf, 40225 Düsseldorf, Germany*
[5]*IMRA America Inc., 1044 Woodridge Avenue, Ann Arbor, Michigan 48105, USA*
[6]*Forschungszentrum Dresden-Rossendorf e. V., Bautzner Landstrasse 128, 01328 Dresden, Germany*
[7]*Queen's University Belfast, Belfast BT7 1NN, United Kingdom*
(Received 26 September 2008; published 26 March 2009)

We report on an electron accelerator based on few-cycle (8 fs full width at half maximum) laser pulses, with only 40 mJ energy per pulse, which constitutes a previously unexplored parameter range in laser-driven electron acceleration. The produced electron spectra are monoenergetic in the tens-of-MeV range and virtually free of low-energy electrons with thermal spectrum. The electron beam has a typical divergence of 5–10 mrad. The accelerator is routinely operated at 10 Hz and constitutes a promising source for several applications. Scalability of the few-cycle driver in repetition rate and energy implies that the present work also represents a step towards user friendly laser-based accelerators.

DOI: 10.1103/PhysRevLett.102.124801
PACS numbers: 41.75.Jv, 41.75.Ht

Laser-driven plasma waves were proposed as compact electron accelerators [1] owing to their ability to produce longitudinal accelerating fields several orders of magnitude larger than those attainable in conventional accelerators. A promising implementation relies on strongly driven "broken" plasma waves. In this regime, the laser intensity is so large that the generated plasma wave breaks directly behind the pulse and some electrons of the background plasma are injected into the first wake of the plasma wave and are accelerated. Numerical studies [2–4] have shown that the accelerated electrons emerge from the plasma as monoenergetic electron bunches with relativistic energy and few-femtosecond duration. Under optimal conditions, the driving laser pulse has relativistic intensity ($> 10^{18}$ W=cm^2) and a duration and diameter that are matched to the plasma density. This demand calls for a pulse length equal to or less than half the plasma wavelength $\lambda_p=2 \approx \pi c = !_p$, and a focal diameter of $\approx \lambda_p$. Here, $!_p = \sqrt{n_e e^2=(\epsilon_0 m_e)}$ is the plasma frequency and e, m_e, n_e stand for the electron charge, mass, and density, respectively. In this case, the ponderomotive force of the laser pulse is so large that the majority of the free electrons are transversally pushed out, leaving the positively charged ions behind. After a propagation length comparable to a plasma wavelength, the electrons are driven back to the axis by the fields built up by charge separation. A cavity void of electrons trailing the laser pulse emerges, it has been dubbed a "bubble" [2]. A fraction of the returning electrons is injected and trapped in the bubble and accelerated by its strong longitudinal electric field, resulting in relativistic electron bunches with narrow-band energy spectra [5,6].

In order to reach the intensities required for this scheme with laser pulses of typical durations between 30–80 fs and matched spot-size, pulse energies in the Joule range are required. This restricts this acceleration mechanism to rather large lasers or requires additional nonlinear self-modulation of the pulse in the plasma, which transforms the laser pulse into the required domain. In the last few years, a number of studies have been devoted to laser-generated monoenergetic electron beams in this regime, starting with the proof of principle in 2004 [7–9] and followed by investigations about the self-modulation of the laser pulse [10], and experiments that increased the energy and/or improved the quality of laser-generated electron beams considerably [11–15].

Here, we demonstrate monoenergetic electron acceleration in a new laser-parameter range by employing for the first time few-cycle laser pulses to this end. Analytical scaling laws [3–5] and our simulations [16] indicate that for laser pulses shorter than 10 fs, this regime can be accessed with less than 100 mJ pulse energy. Thanks to a laser pulse duration of 8 fs, a laser pulse energy as low as 40 mJ enabled the acceleration process to work, giving rise to clean, monoenergetic electron spectra in the range of several 10 MeV.

In our experiments, we used LWS-10 (Light Wave Synthesizer-10), the world's first multi-TW sub-10-fs light source [17]. It draws on a noncollinear optical parametric chirped pulse amplifier, allowing the amplification of broad bandwidth pulses. In our investigations, the system produced pulses with 50-mJ energy, 8 fs duration, and spectra covering the range of 700–980 nm at a 10-Hz repetition rate. Because of losses in the beam line connect-

0031-9007=09=102(12)=124801(4) 124801-1 © 2009 The American Physical Society

ing laser and experiment, the energy on target was reduced to 40 mJ.

The laser pulses are focused by a gold-coated f=6 off-axis parabola onto a helium gas jet (see Fig. 1) to a spot diameter of 6 μm FWHM, yielding an on-axis peak intensity of 1.2×10^{19} W=cm². An adaptive mirror in closed loop mode is used to correct wave front aberrations.

The helium interaction medium is provided either by a supersonic gas jet with an approximately constant density within a diameter of 300 μm or by a gas jet with a Gaussian-shaped density profile produced by a subsonic nozzle with a diameter of 400 μm. The gas density can be varied between 10^{18} and 10^{20} cm⁻³. The laser-generated plasma channel is imaged transversally by a long-object-distance microscope objective onto a CCD camera, allowing high-resolution measurements of channel diameter and length. In addition, a weak probe beam can be coupled into the side-view imaging system for assisting in the alignment of the nozzle and for studying the plasma channel.

The electron energy spectrum is measured by a high-resolution focusing permanent-magnet spectrometer suitable for analyzing electrons in the range of 2–400 MeV. It comprises a 30-cm × 40-cm focusing permanent magnet with a magnetic field of 1 T over a gap of 5 cm and uses 600 highly sensitive scintillating fibers coupled to a 16-bit CCD camera. Alternatively, a scintillating screen (Kodak Lanex) is imaged to a 12-bit CCD camera, allowing simultaneous measurement of energy spectrum and divergence. Lanex and fibres were cross-calibrated using an image plate [18], but subsequent changes in the detection system restricted the accuracy of the bunch-charge measurement to a factor of 2. Low-energy electrons down to ~100 keV were detected and analyzed by a smaller spectrometer.

For both types of gas jets used in our experiments, optimal conditions for the accelerator were obtained with an electron density of 2×10^{19} cm⁻³. This matches very well the requirements of the bubble regime for our laser pulse duration and focusing conditions.

Typical spectra of electron beams emerging from the plasma accelerator using the 400 μm cylindrical nozzle are shown in Fig. 2. The displayed spectra comprise a very low number of thermal electrons at low energies and pronounced monoenergetic peaks between 13 and 23 MeV comprising a charge of the order of 10 pC.

Figure 3 shows raw data and calibrated line outs of a shot captured by the scintillating screen. The spectrum shows a monoenergetic peak at 24.6 MeV with 3.3% relative energy spread (FWHM), a transversal beam divergence of 6.3 mrad (FWHM) and a charge of approximately 3 pC. Two weak features at higher energies are also discernible; they probably stem from (weak) injection and acceleration in different plasma wave troughs. This effect was also observed in our simulations.

The side-view imaging of the interaction region revealed a weakly scattering, channel-like structure with a length of approximately 100–120 μm, which is approximately equal to the Rayleigh range of the laser beam, and a diameter of approximately 5 μm. Strong scattering, indicative of self-modulation observed in experiments with longer laser pulses, was not observed. The measured channel length and electron energies of 10–30 MeV imply an average longitudinal accelerating gradient of 0.1–0.3 TeV=m, assuming acceleration taking place over the entire channel length. Since no pronounced structure of the channel was observed, the point of injection could not be determined, but measurements with a gas jet of only 150 μm diameter led to similar electron beam parameters as the larger jets, providing an upper limit for the acceleration length. The above mentioned gradient is also in

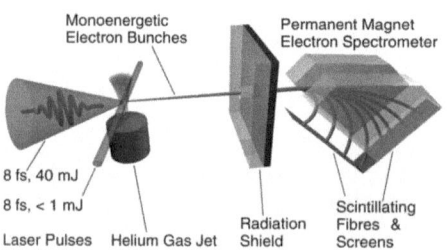

FIG. 1 (color). Schematic illustration of the experiment. The 8-fs laser pulses are split into a driver beam with 40 mJ energy and a weak probe beam with less than 1 mJ energy. The driver beam is focused onto a helium jet with electron densities between 10^{18} to 10^{20} cm⁻³. The electron beam emerging from the laser-plasma interaction is characterized by a permanent-magnet electron spectrometer.

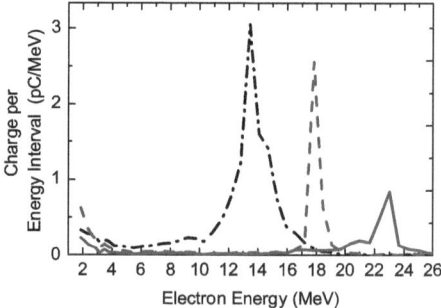

FIG. 2. Typical spectra of monoenergetic electron beams from the sub-10-fs laser-driven accelerator. They have mean energies of 13.4, 17.8, and 23 MeV. The bunches carry a charge of approximately 10, 3.5, and 1.6 pC, respectively. All three spectra show remarkably few thermal background electrons, an observation that has been confirmed down to energies of 100 keV using a smaller spectrometer. These spectra were obtained with the cylindrical (subsonic) 400 μm gas jet.

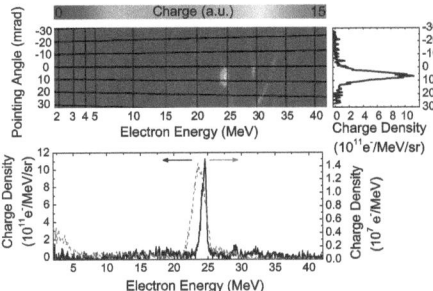

FIG. 3 (color). Electron spectrum and transverse beam size obtained with a scintillating screen placed in the electron spectrometer and imaged onto a CCD camera. The false-color plot shows the image on the screen with the right and the bottom panels depicting calibrated line outs along the energy axis (lower plot) and the transversal axis (right side). The displayed result was obtained with a 300-μm supersonic gas jet and exhibits a monoenergetic peak at 24.6 MeV with 3.3% energy spread (FWHM), a divergence of 6.3 mrad (FWHM), and a charge of 3 pC. Also shown is the simulated electron spectrum (red, dashed line), which is in reasonable agreement with the experimental result.

good agreement with values previously reported [19] as well as with our simulations.

An important feature of the observed electron spectra is the apparently very low number of low-energy electrons. These "thermal" electrons typically show up in the spectrum of laser-accelerated electrons with an exponentially decaying energy distribution in the energy range between 1 to 10 MeV [19–21]. We confirmed the low number of thermal electrons down to energies of 100 keV by employing an electron spectrometer optimized for the energy range of 100 keV to 13 MeV. Quantitative comparison with earlier experiments is hampered by the lack of spectral analysis of the thermal electrons in most previous studies. The low thermal background also leads to a low dose of gamma radiation generated by the electron beam, which—together with low gas load due to the small gas jets—allowed us to routinely operate the accelerator at a repetition rate of 10 Hz.

To gain more insight into the details of the acceleration process, we performed numerical simulations with the 3-dimensional particle-in-cell (3D PIC) code ILLUMINATION [16]. The plasma is modeled with a uniform transversal density distribution, whereas in the longitudinal direction, a 120-μm broad flattop profile, matching the experimentally determined channel length, with an electron density of 2×10^{19} cm^{-3} terminated by exponential gradients, is assumed. The 1=e scale length of the entrance gradient is chosen to be 5 μm, to avoid numerical problems with a steep gradient. The exit gradient was chosen as 30 μm to match the experimental conditions. The laser pulse had a

(FWHM) duration of 8.5 fs, a Gaussian beam waist of 6 μm, a pulse energy of 38 mJ, and a carrier wavelength of 800 nm.

Figure 4 shows a snapshot of the electron density distribution, laser intensity, and longitudinal electrical field of the plasma accelerator after the laser pulse has travelled 125 μm across the 155-μm-thick gas jet. The strongly driven plasma wave, forming electron voids ("bubbles") trailing the laser pulse, as well as wave breaking and injection are conspicuous. Figure 4(b) shows line outs of on-axis instantaneous laser intensity and electron density along the optical axis of the laser beam in a propagation

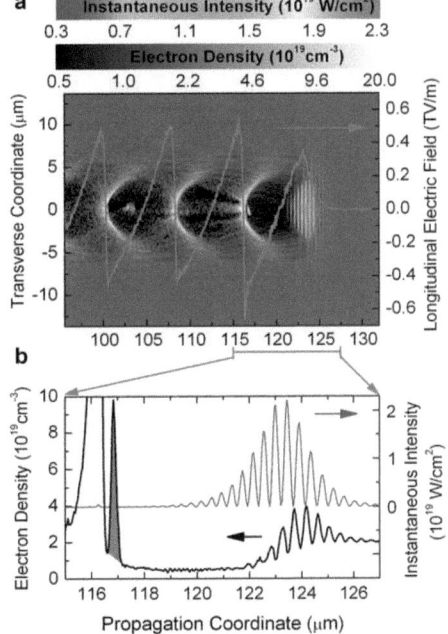

FIG. 4 (color). Simulation results showing the physical state of the system after the laser pulse has travelled a distance of 125 μm inside the fully preionized plasma. Panel (a) shows the electron density (grey-scale plot) and the instantaneous laser intensity (rainbow-scale false-color diagram). The laser intensity plot is clipped along the contour line where the intensity drops by a factor of 1=e^2 with respect to its peak value. The red line shows the longitudinal accelerating field caused by charge separation in the bubble, amounting to about 0.45 TV=m at the bunch location. Panel (b) plots line outs of electron density and instantaneous laser intensity along the optical axis of the laser beam. The injected electron bunch (red) contains 4.5 pC charge and has almost an order of magnitude higher density than the background plasma.

interval confining the laser pulse and the first bubble behind it. The laser pulse expels approximately 75% of the free electrons in the vicinity of the propagation axis. The expelled electrons swing back to the axis forming the prominent density peak at the back of the bubble. Some of the returning electrons are then scattered into the bubble by the space charge field of that peak and accumulate within the bubble, forming a dense electron bunch (shown in red) that is trapped and accelerated. The accelerating longitudinal electrical field is produced by the density peak at the bubble vertex and by the positively charged ions inside the bubble, Fig. 4(a) (red line). It is hardly perturbed by the space charge of the bunch itself, in principle permitting further loading of the beam. The electron spectrum after the electron bunch has left the plasma and propagated several 100 μm into the surrounding vacuum is given in Fig. 3. It shows a monoenergetic peak at 24 MeV, which implies together with the accelerating field at the bunch location of 0.45 TV=m an effective acceleration length of about 60 μm. The electron bunch emerging from the plasma carries a charge of about 4.5 pC, has a duration of about 1 fs, and is accompanied by a small exponential background. The ultrashort electron bunch duration is approximately preserved within the range of simulated propagation extending several 100 micrometers behind the gas jet.

Our simulations show that our current laser parameters are close to the threshold for self-injection and formation of a stable accelerating structure [16]. As can be seen in Fig. 4(a), the laser intensity is just high enough to produce self-injection but still keep the wave breaking small enough to allow several oscillations of the wake wave. This is in good agreement with both the analytic theory of bubble acceleration [5], which—for our pulse duration—predicts the onset of the process at pulse energies of about 30 mJ, and the experimental finding that gradual reduction of the laser pulse energy results in rapid increase of fluctuations of the electron beam properties, with acceleration ceasing completely for pulse energies below 25 mJ. Similarly, a deviation of the plasma density of 10% from the optimum value reduced the yield of monoenergetic spectra by as much as a factor of 2. Also stretching the pulses to more than 13–14 fs lowered the yield of monoenergetic spectra dramatically. This implies that with our current on-target energy of 40 mJ, we operate the bubble accelerator close to the boundary of its operational regime. Higher driving pulse energies should not only improve stability but—according to our simulations—also dramatically increase the laser-to-electron energy conversion efficiency, from currently $\sim$1% to up to 20%. Furthermore, the simulations also predict that the laser pulse is rapidly depleted by energy transfer to the background plasma and therefore looses its ability to sustain a stable bubble already after 250 μm propagation. To verify this, an extremely small Laval nozzle with only 150 μm exit diameter was used to produce a gas target with a length shorter than the depletion length of the laser, which led to an improved stability but also to somewhat lower electron energies.

Our experimental and numerical results make few-cycle-laser-driven electron acceleration appear a promising concept for producing relativistic, monoenergetic electron bunches with ultrashort duration. With 8 fs laser pulses containing 40 mJ energy, we generated monoenergetic electron bunches with multi-10 MeV energy, approximately 10 pC charge and 5 to 10 mrad (FWHM) divergence. The electron accelerator is routinely operated at 10 Hz, which renders this electron accelerator a promising source for many applications including time-resolved laser-pump electron-probe experiments and seeding of wake-field acceleration stages. The reliance on optical parametric amplification in the generation of the laser driver offers the potential of increasing the repetition rate and pulse energy by orders of magnitude. Hence, this study also constitutes a first step towards the development of user friendly laser-based electron accelerators.

We would like to thank Stefan Karsch for many helpful discussions. This work was supported by DFG-Project Transregio TR18 by the Association EURATOM—Max-Planck-Institut fuer Plasmaphysik and by The Munich Centre for Advanced Photonics (MAP).

---

*karl.schmid@mpq.mpg.de; laszlo.veisz@mpq.mpg.de
[1] T. Tajima and J. M. Dawson, Phys. Rev. Lett. **43**, 267 (1979).
[2] A. Pukhov and J. Meyer-Ter-Vehn, Appl. Phys. B **74**, 355 (2002).
[3] W. Lu et al., Phys. Rev. Lett. **96**, 165002 (2006).
[4] F. S. Tsung et al., Phys. Plasmas **13**, 056708 (2006).
[5] S. Gordienko and A. Pukhov, Phys. Plasmas **12**, 043109 (2005).
[6] A. Pukhov and S. Gordienko, Phil. Trans. R. Soc. A **364**, 623 (2006).
[7] C. G. R. Geddes et al., Nature (London) **431**, 538 (2004).
[8] S. P. D. Mangles et al., Nature (London) **431**, 535 (2004).
[9] J. Faure et al., Nature (London) **431**, 541 (2004).
[10] B. Hidding et al., Phys. Rev. Lett. **96**, 105004 (2006).
[11] W. P. Leemans et al., Nature Phys. **2**, 696 (2006).
[12] S. Karsch et al., New J. Phys. **9**, 415 (2007).
[13] T. P. Rowlands-Rees et al., Phys. Rev. Lett. **100**, 105005 (2008).
[14] J. Faure et al., Nature (London) **444**, 737 (2006).
[15] J. Osterhoff et al., Phys. Rev. Lett. **101**, 085002 (2008).
[16] M. Geissler, J. Schreiber, and J. Meyer-Ter-Vehn, New J. Phys. **8**, 186 (2006).
[17] F. Tavella et al., Opt. Lett. **32**, 2227 (2007).
[18] B. Hidding et al., Rev. Sci. Instrum. **78**, 083301 (2007).
[19] C.-T. Hsieh et al., Phys. Rev. Lett. **96**, 095001 (2006).
[20] A. Yamazaki et al., Phys. Plasmas **12**, 093101 (2005).
[21] S. Masuda et al., Phys. Plasmas **14**, 023103 (2007).

Appendix B9

Density-transition based electron injector for laser driven wakefield accelerators
K. Schmid, A. Buck, C. M. S. Sears, J. M. Mikhailova, R. Tautz, D. Herrmann, M. Geissler, F. Krausz, and L. Veisz

Reprinted with permission from
Physical Review Special Topics-Accelerators and Beams 13, 091301 (2010).

DOI: 10.1103/PhysRevSTAB.13.091301

http://link.aps.org/doi/10.1103/PhysRevSTAB.13.091301

Copyright © 2011, American Physical Society.

Density-transition based electron injector for laser driven wakefield accelerators

K. Schmid,[1,*] A. Buck,[1] C. M. S. Sears,[1] J. M. Mikhailova,[1,2] R. Tautz,[1,3] D. Herrmann,[1,4] M. Geissler,[5] F. Krausz,[1,6] and L. Veisz[1,†]

[1]*Max-Planck-Institut für Quantenoptik, Hans-Kopfermann-Strasse 1, 85748 Garching, Germany*
[2]*A.M. Prokhorov General Physics Institute, Russian Academy of Science, Moscow, Russia*
[3]*LS für Photonik und Optoelektronik, LMU München, Amalienstrasse 54, 80799 München, Germany*
[4]*LS für BioMolekulare Optik, LMU München, Oettingenstrasse 67, 80538 München, Germany*
[5]*Queen's University Belfast, Belfast BT7 1NN, United Kingdom*
[6]*Ludwig-Maximilians-Universität München, Am Coulombwall 1, 85748 Garching, Germany*
(Received 5 July 2010; published 7 September 2010)

We demonstrate a laser wakefield accelerator with a novel electron injection scheme resulting in enhanced stability, reproducibility, and ease of use. In order to inject electrons into the accelerating phase of the plasma wave, a sharp downward density transition is employed. Prior to ionization by the laser pulse this transition is formed by a shock front induced by a knife edge inserted into a supersonic gas jet. With laser pulses of 8 fs duration and with only 65 mJ energy on target, the accelerator produces a monoenergetic electron beam with tunable energy between 15 and 25 MeV and on average 3.3 pC charge per electron bunch. The shock-front injector is a simple and powerful new tool to enhance the reproducibility of laser-driven electron accelerators, is easily adapted to different laser parameters, and should therefore allow scaling to the energy range of several hundred MeV.

DOI: 10.1103/PhysRevSTAB.13.091301   PACS numbers: 41.75.Ht, 42.65.Jx, 42.65.Re, 42.65.Yj

Laser wakefield accelerators [1] are intensely investigated due to their ability to produce longitudinal accelerating gradients several orders of magnitude larger than those attainable in conventional accelerators. Especially since the first experimental evidence of monoenergetic electron spectra produced by such accelerators [2–4] interest in this field multiplied, culminating in the generation of monoenergetic electron beams with 1 GeV energy [5]. All these experiments relied on self-injection of plasma electrons into a laser-driven plasma wave by wave breaking [6]. In this regime the intensity of the driving laser pulse is so large that the plasma wave breaks and some electrons of the background plasma are injected into the first wake of the plasma wave. The longitudinal electric field built up by the plasma wave rapidly boosts the speed of these injected electrons close to the velocity of light. They become trapped and are subsequently further accelerated to highly relativistic energies by the wave [7]. This self-injection offers the great benefit of obviating the need for the experimentally very challenging injection of an externally generated electron pulse. The inherent disadvantage of this injection technique is posed by the absence of control over the exact locations at which wave breaking—and hence injection—starts and stops and over the amount of injected charge. In order to gain some control over the injection process several different approaches have been adopted so far: injection at a density down ramp produced electrons with a momentum of 0.76 ± 0.02 MeV=c [8], enhanced injection was also shown using mixtures of different gases [9,10]. A scheme employing a counterpropagating laser pulse [11,12] was shown to improve the reproducibility of the generated electron beam.

Here we present a simple scheme that allows for precisely localized injection of electrons into a laser-driven plasma wave and subsequent acceleration yielding a monoenergetic spectrum. This scheme draws on a sharp downward (along laser propagation) plasma density transition between two adjacent regions of different densities [13]. The laser intensity and plasma density are adjusted such that the interaction is highly nonlinear but no wave breaking occurs while the laser propagates in either of the two regions. This allows for the generation of a stable but highly anharmonic plasma wave, which, in turn, provides large accelerating gradients. During the downward density transition, the plasma wavelength λ_p increases abruptly from its high-density value to the low-density one. Here, the plasma wavelength is given by $\lambda_p \approx 2\pi c / \omega_p$ and the plasma frequency by $\omega_p = \sqrt{n_e e^2/(\epsilon_0 m_e)}$, with e, m_e, n_e being electron charge, mass, and density, respectively. This sudden increase in plasma wavelength causes a rephasing of a sizable fraction of the plasma electrons into the accelerating phase of the plasma wave. We point out that this is different to the injection caused during density downramps that extend over many plasma periods [8], where a reduced plasma wave phase velocity enables effi-

*karl.schmid@mpq.mpg.de
†laszlo.veisz@mpq.mpg.de

cient trapping of nonrelativistic plasma electrons. In contrast, in the case of the sharp transition, the plasma wave is fully loaded at once due to the sudden increase in plasma wavelength. Electron trapping at sharp downward density transitions has been extensively studied theoretically [14–16]. *Sharp* in this context means that the characteristic length of the transition is on the order of the plasma wavelength [17]. Up to now experimental realizations of this scheme did not produce monoenergetic electron beams and relied on a second laser beam that induces a plasma density transition via local plasma heating [18–20].

In contrast to former schemes, in the present experimental approach the density transition is formed in the gas jet *prior* to ionization by the driving laser pulse. It exploits shock-front formation in a supersonic He-gas jet. The jet is generated by a pulsed de Laval nozzle with an exit diameter of 300 μm; the shock is generated by introducing a knife edge laterally into the gas jet. In contrast to a subsonic flow, the supersonic flow cannot adapt upstream to the obstacle and therefore needs to adapt locally in the form of an abrupt change of all flow parameters. The typical length scale of this sudden change is on the order of a few times the molecular mean free path depending on initial Mach number and shock angle [21,22]. The (asymptotic) values of the flow parameters before and after the shock can be calculated using the Euler equations. For the ratio of gas densities in front of and behind the shock, one obtains [23]

$$\frac{n_1}{n_2} = 1 - \frac{2}{\kappa + 1}\left[1 - \frac{1}{(M_1 \sin\alpha)^2}\right] ; \quad (1)$$

where n_1 and n_2 are the gas densities before and after the shock, respectively, κ is the specific heat ratio with a value of 5=3 for a monoatomic gas, M_1 is the initial Mach number of the gas flow, and α (see Fig. 1) is the angle between the gas flow and the shock front. Relation (1) exhibits a minimum for $\alpha = 90°$ corresponding to a strong perpendicular shock and a maximum value of 1 for $\alpha_m = \arcsin(1=M_1)$ for weak distortions propagating at the Mach angle. A sketch of the nozzle setup is given in Fig. 1(a). The razor blade is mounted on a small translation stage allowing to switch the shock front on and off as well as moving the shock transversally through the gas jet. Since moving the knife edge through the gas jet not only changes the shock-front position but also the angle, the density ratio changes accordingly.

The laser employed in the present experiments is the multi-TW sub-10-fs light source light wave synthesizer 20 (LWS-20) [24], an upgraded version of the system used for the experiments in [25]. During the experiments it delivered pulses with 65 mJ energy on target and a duration of 8 fs full width at half maximum (FWHM). The laser pulses are focused onto the target by a f=12 off-axis parabolic mirror (OAP) to a spot diameter of 12 μm (FWHM),

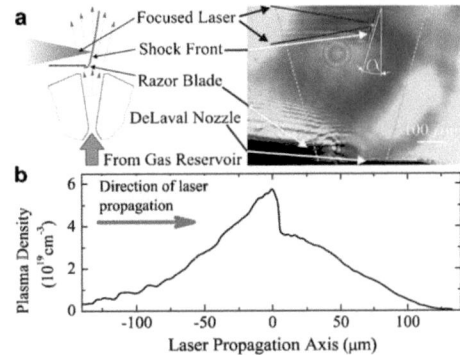

FIG. 1. Sketch and shadowgraph image of the target setup (a). A razor blade is introduced laterally into a supersonic gas jet (undistorted gas jet edges: white dashed lines) in order to generate a shock front (blue line). The shadowgraph image was taken during the experiments showing the nozzle tip, the razor blade (slightly tilted), and the plasma produced by the driving laser. Inside the plasma the tilted shock front can be discerned and a small, bright spot indicative of electron injection is visible. (b) Measured electron density along the laser propagation axis.

yielding a peak intensity of 2.5×10^{18} W=cm^2. This value takes into account losses introduced by residual phase front distortions.

The electron energy spectrum is measured by a high resolution focusing permanent magnet spectrometer [26] suitable for analyzing electrons in the range of 2–400 MeV. Scintillating screens (Kodak Lanex) imaged to a 12 bit CCD camera are used for electron detection, allowing simultaneous measurement of energy spectrum and divergence. This detection system was absolutely calibrated at a linear accelerator source [27].

Shadowgraph images of the plasma [Fig. 1(a)] are obtained during the experiments by using a small part of the laser beam as back light. By introducing a Nomarski interferometer [28] into this probe beam, the same back light is used for interferometric measurements of the plasma density. Lineouts parallel to the shock front are taken out of the phase-shift maps produced by the measurement and subsequent Abel inversion yields the plasma density distribution. It was confirmed that the tilt by the angle α of the lineouts with respect to the plasma channel does not significantly change the results. A density lineout along the laser axis is shown in Fig. 1(b).

Figures 1(a) and 1(b) yield $\alpha = 16 \pm 0.5°$ and $n_1=n_2 ' 1.6$. Substitution into formula (1) gives a Mach number $M = 5.1 \pm 0.2$ which is in good agreement with $M = 5.3$ obtained from computational fluid dynamics simulations.

TABLE I. Comparison self-injection and shock-front injection.

Parameter	Self-injection	Density transition
Energy (MeV)	$26.0^{+8.2}_{-6.6}$	$23.3^{+3.8}_{-3.0}$
Energy spread (%)	12^{+9}_{-10}	9^{+6}_{-8}
Divergence (mrad)	$10.9^{+3.5}_{-3.7}$	$8.9^{+3.1}_{-3.3}$
Charge (pC)	$3.7^{+2.9}_{-3.4}$	$3.3^{+2.0}_{-2.2}$

For the resultant effective Mach number $M_1 \sin(\alpha)$ of 1.4, the width of the hydrodynamic shock is expected to be roughly 1 order of magnitude larger than the molecular mean free path [22] which for He at a neutral gas density of 1.9×10^{19} cm^{-3} is approximately 0.6 μm. From Fig. 1 we evaluate the width of the density transition as $\approx$ 5 μm, in agreement with this expectation.

After fine-tuning the plasma density and focal spot position, quasimonoenergetic electron energy spectra are obtained. As in [25], the spectra are virtually background-free. Depending on the day of operation, the occurrence probability of a quasimonoenergetic electron spectrum varies between 50% and 95%. In order to quantify the improvement obtained with this new injection scheme, Table I shows a comparison of electron beam parameters produced by self-injection and by shock-front injection. Errors are root-mean-square (RMS) deviations from the average value. To account for asymmetric distributions, the RMS value was independently calculated for the occurrences with higher-than-average and lower-than-average value. In Fig. 2 and Table I, energy spread and divergence refer to FWHM.

It can be seen that for density-transition injection, all shot-to-shot fluctuations as well as the energy spread and divergence are smaller than in the self-injection case. In both cases, approximately 36% of all the shots fall within the RMS error intervals with respect to all four parameters simultaneously. Since these error intervals are much smaller for the density-transition case, this again shows a clear advantage of injection at sharp density transitions.

Representative electron spectra out of the 10% fraction with lowest energy spread are displayed in Fig. 2. It is evident that the shock-front injection scheme yields a much cleaner and more reproducible beam than obtainable with self-injection. The accelerator is routinely operated at the laser repetition rate of 10 Hz; therefore, the electron accelerator currently produces approximately one high quality shot per second on a daily basis and is, thus, ready for first applications. To support the claim that electron injection is indeed happening exclusively at the location of the shock front, we note first that at the crossing point between the laser beam and the shock front a bright spot of broadband light emission occurs in the side view images [see Fig. 1(a)] resembling the wave-breaking radiation reported in [29]. It was confirmed that the brightness of the spot is correlated to the accelerated charge. Further support is provided by the fact that the electron energy scales linearly with the position of the shock front and, hence, with the acceleration length. In this way the electron energy could be varied between 15 and 23 MeV; the accelerating gradient was determined as $\sim$190 GV=m. This result is presented in Fig. 3.

Numerical simulations with the three-dimensional particle-in-cell (3D-PIC) code ILLUMINATION [30] pro-

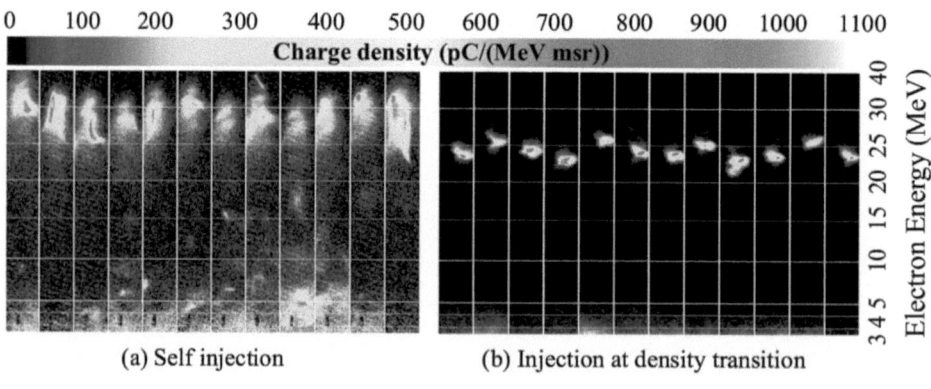

FIG. 2. A few shots representative for those 10% of all the shots with lowest energy spread for self-injection (a) and injection at a density transition (b). The horizontal axis in each image corresponds to the transversal electron beam size; the vertical axis shows electron energy. For self-injection the parameters of this fraction are 29.7 $\pm$ 1.2 MeV energy, 8 $\pm$ 5% energy spread, 10.0 $\pm$ 2.3 mrad divergence, and 3.1 $\pm$ 1.0 pC charge. For injection at the density transition, these values are 24.3 $\pm$ 0.9 MeV, 4 $\pm$ 0.5%, 7.3 $\pm$ 0.5 mrad, and 1.8 $\pm$ 0.5 pC, respectively.

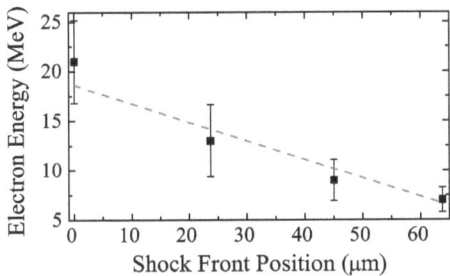

FIG. 3. Variation of electron energy with the position of the density transition. The grey dashed line shows a linear fit yielding an acceleration gradient ′ 190 GV=m.

vided further details of the injection process. The laser pulse and plasma parameters used in the simulation match those of the experiment. Figure 4 shows, from top to bottom, the physical state of the laser-driven plasma wave shortly before, immediately after and ~85 fs after the laser pulse crossed the downward density transition. The plasma wave is strongly driven by the laser pulse, but does not break until the laser pulse crosses the density transition. At the transition the wave breaks and a compact electron bunch with 60 pC charge is injected into the first wave bucket after the laser pulse, leading to a complete destruction of the following plasma wave buckets. This is to say, the wave fully breaks down at this point, leaving only the accelerating bubble structure. The accelerating field at the position of the electron bunch is 250 GV=m, in reasonable agreement with the experimental value. The PIC simulations confirm that electrons are exclusively injected at the density transition and show the injection mechanism to be quite robust against changes of the upper and lower density levels, a fact which was also confirmed experimentally.

In conclusion, we present a simple, reliable, and robust method of injecting background electrons into a laser-driven plasma wake. It relies on injection at a sharp downward density transition originating from a shock front which is generated by a razor blade introduced laterally into a supersonic gas jet. The resulting electron beam is typically monoenergetic and has a significantly improved stability and reduced energy spread. Additionally, the electron energy can be tuned in a wide range. It is important to note that the present setup is easily adapted to different laser parameters and—once applied to multi-Joule laser pulses—could pave the way towards a reliable tabletop electron accelerator in the multi-100 MeV range. By generating a supersonic shock front inside a guiding structure like a capillary [5], even the GeV energy range comes into reach.

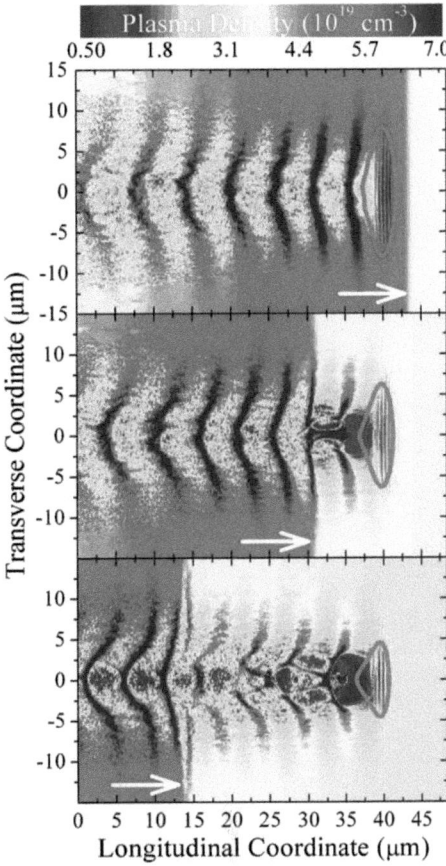

FIG. 4. State of the accelerator as obtained from PIC simulations shortly before (upper panel), immediately after (middle panel), and ~85 fs after (lower panel) the laser pulse has crossed the density transition. White arrows indicate the position of the density transition; grey lines show the contour where 13.5% of maximum laser intensity is reached. The normalized instantaneous laser intensity is shown in a rainbow color map. The upper panel shows a strongly driven plasma wave that does not break. Strong injection triggered by the shock front is evident in the middle panel. The strong electric field of the injected electron bunch destroys the remaining plasma wave thus impeding injection in more than one wave-trough. This leads to a fully developed bubble structure (lower panel) that accelerates the injected electrons to several 10 MeV energy. Since the bubble is fully loaded with electrons, no further injection occurs.

ACKNOWLEDGMENTS

This work is supported by DFG-Project Transregio TR18, by the Association EURATOM—Max-Planck-Institut fuer Plasmaphysik, by The Munich Centre for Advanced Photonics (MAP) and by the Laserlab-Europe/ Labtech FP7 Contract No. 228334. C. M. S. Sears acknowledges the support of the Alexander von Humbold Foundation. J. M. Mikhailova acknowledges support by Alexander-von-Humboldt Foundation and RFBR, Grants No. 08-02-01245-a and No. 08-02-01137-a. D. Herrmann is grateful to Studienstiftung des deutschen Volkes.

[1] T. Tajima et al., Phys. Rev. Lett. **43**, 267 (1979).
[2] C. Geddes et al., Nature (London) **431**, 538 (2004).
[3] S. Mangles et al., Nature (London) **431**, 535 (2004).
[4] J. Faure et al., Nature (London) **431**, 541 (2004).
[5] W. P. Leemans et al., Nature Phys. **2**, 696 (2006).
[6] S. V. Bulanov et al., Phys. Rev. Lett. **78**, 4205 (1997).
[7] A. Pukhov et al., Appl. Phys. B **74**, 355 (2002).
[8] C. Geddes et al., Phys. Rev. Lett. **100**, 215004 (2008).
[9] A. Pak et al., Phys. Rev. Lett. **104**, 025003 (2010).
[10] C. McGuffey et al., Phys. Rev. Lett. **104**, 025004 (2010).
[11] J. Faure et al., Nature (London) **444**, 737 (2006).
[12] C. Rechatin et al., Phys. Rev. Lett. **102**, 164801 (2009).
[13] S. Bulanov et al., Phys. Rev. E **58**, R5257 (1998).
[14] H. Suk et al., Phys. Rev. Lett. **86**, 1011 (2001).
[15] P. Tomassini et al., Phys. Rev. ST Accel. Beams **6**, 121301 (2003).
[16] H. Suk et al., J. Opt. Soc. Am. B **21**, 1391 (2004).
[17] A. V. Brantov et al., Phys. Plasmas **15**, 073111 (2008).
[18] J. Kim et al., Phys. Rev. E **69**, 026409 (2004).
[19] T.-Y. Chien et al., Phys. Rev. Lett. **94**, 115003 (2005).
[20] J. Kim et al., J. Korean Phys. Soc. **51**, 397 (2007).
[21] D. Haenel, *Molekulare Gasdynamik* (Springer-Verlag, Berlin, 2004), 1st ed.
[22] H. M. Mott-Smith, Phys. Rev. **82**, 885 (1951).
[23] R. D. Zucker and O. Biblarz, *Fundamentals of Gas Dynamics* (Wiley, New York, 2002).
[24] D. Herrmann et al., Opt. Lett. **34**, 2459 (2009).
[25] K. Schmid et al., Phys. Rev. Lett. **102**, 124801 (2009).
[26] C. M. S. Sears et al. (to be published).
[27] A. Buck et al., Rev. Sci. Instrum. **81**, 033301 (2010).
[28] R. Benattar et al., Rev. Sci. Instrum. **50**, 1583 (1979).
[29] A. Thomas et al., Phys. Rev. Lett. **98**, 054802 (2007).
[30] M. Geissler et al., New J. Phys. **8**, 186 (2006).

Appendix B10

Toward single attosecond pulses using harmonic emission from solid-density plasmas
*P. Heissler, R. Hörlein, M. Stafe, J.M. Mikhailova, Y. Nomura, D. Herrmann,
R. Tautz, S.G. Rykovanov, I.B. Földes, K. Varjú, F. Tavella, A. Marcinkevicius,
F. Krausz, L. Veisz, G.D. Tsakiris*

Reprinted with permission from
Applied Physics B 101, 511–521 (2010).

DOI: 10.1007/s00340-010-4281-6

http://www.springerlink.com/content/qn854754v7k161x6/

Copyright © 2011, Springer.

Toward single attosecond pulses using harmonic emission from solid-density plasmas

P. Heissler · R. Hörlein · M. Stafe · J.M. Mikhailova · Y. Nomura · D. Herrmann ·
R. Tautz · S.G. Rykovanov · I.B. Földes · K. Varjú · F. Tavella · A. Marcinkevicius ·
F. Krausz · L. Veisz · G.D. Tsakiris

Received: 16 July 2010 / Revised version: 20 September 2010 / Published online: 30 October 2010
© Springer-Verlag 2010

Abstract We report on investigations of high-order harmonic generation from solid surfaces in the coherent wake emission regime with relativistically intense few-cycle (8 fs) laser pulses. Significant spectral broadening compared to previous experiments with many-cycle pulses and the appearance of substructures on the harmonics are observed that strongly fluctuate from shot-to-shot. Measurements in which the linear polarization was rotated or ellipticity of the laser pulse was varied exhibit a strong dependence of the harmonic emission on the polarization state of the incident pulse. We show that the observed spectral features are ultimately connected to the sub-cycle electron dynamics in the laser-solid interaction and thus proof of the few-cycle nature of the observed harmonic emission. Using a simple model we have investigated the factors that play an important role in the shape of the emitted spectrum.

1 Introduction

It has been recently demonstrated that the harmonic emission emanating from the interaction of intense laser pulses with solid-density plasma possesses a remarkable property, the individual harmonics in the emission spectrum are phase-locked [1, 2]. This conclusion is based on the experimental observation that in the time domain the emitted

P. Heissler (✉) · R. Hörlein · J.M. Mikhailova · Y. Nomura ·
D. Herrmann · R. Tautz · S.G. Rykovanov · F. Tavella ·
A. Marcinkevicius · F. Krausz · L. Veisz · G.D. Tsakiris
Max-Planck-Institut für Quantenoptik, 85748 Garching, Germany
e-mail: patrick.heissler@mpq.mpg.de

R. Hörlein · F. Krausz
Fakultät für Physik, Ludwig-Maximilians-Universität München, 85748 Garching, Germany

M. Stafe
Department of Physics, University "Politehnica" of Bucharest, 060042 Bucharest, Romania

J.M. Mikhailova
A.M. Prokhorov General Physics Institute, Russian Academy of Science, 119991 Moscow, Russia

S.G. Rykovanov
Moscow Engineering Physics Institute, 115409 Moscow, Russia

I.B. Földes
KFKI-Research Institute for Particle and Nuclear Physics, 1525 Budapest, Hungary

K. Varjú
Department of Optics and Quantum Electronics, University of Szeged, 6720 Szeged, Hungary

Present address:
Y. Nomura
Institute for Solid State Physics, University of Tokyo, Kashiwa, Chiba 277-8581, Japan

Present address:
D. Herrmann
Lehrstuhl für BioMolekulare Optik, Ludwig-Maximilians-Universität München, 80538 München, Germany

Present address:
R. Tautz
Department of Physics and CeNS, Photonics and Optoelectronics Group, Ludwig-Maximilians-Universität München, 80799 München, Germany

Present address:
F. Tavella
DESY, 22603 Hamburg, Germany

Present address:
A. Marcinkevicius
IMRA America Inc., Ann Arbor, MI 48105, USA

light is bunched in form of a train of sub-laser-cycle duration pulses. Moreover, substantial theoretical [3–8] and experimental [9–15] evidence has been accumulated indicating that the plasma medium additionally exhibits the characteristics required for a source of spatially coherent, extreme ultraviolet (XUV) pulses of attosecond duration and unprecedented brightness. This new source is superior to the one based on the harmonic emission from gaseous media in several aspects. Besides being more efficient, it exhibits no inherent limitation on the laser intensity that can be used. In conjunction with rapid progress in laser technology, the plasma medium holds promise for building a source delivering attosecond pulses [16] with enough intensity to be used in XUV-pump XUV-probe spectroscopy. For most of the envisaged applications, the availability of a single attosecond pulse is desirable [17]. There are two approaches toward this objective. Whereas the first relies on the "intensity gating" technique that requires modern laser systems delivering pulses with some Joules of energy and few-cycle duration [6], the second is based on the "polarization gating" technique [18–20] that can be used with the more readily available conventional laser systems delivering 20–40 fs laser pulses. In both cases, Carrier-Envelope-Phase (CEP) stabilized pulses are necessary [21].

In this work, we report on investigations pertaining to both approaches. Using the high-power Light Wave Synthesizer (LWS) [22, 23] delivering 3-cycle (8 fs), non-phase-stabilized 800 nm, 16 TW laser pulses, we were able to observe for the first time high harmonics up to the 20th order (H20) generated by the interaction of few-cycle laser pulses with solid targets. During these investigations, the parameters associated with the laser pulse from the LWS laser system were such that the dominant harmonic emission mechanism appears to be what by now is dubbed as Coherent Wake Emission (CWE) [9, 24]. Generally, CWE prevails for values of the normalized vector potential up to $a_L \sim 1$, which in terms of the focused laser intensity I_L and wavelength λ_L is given by $a_L^2 = I_L \lambda_L^2 / [1.38 \times 10^{18}\ \text{W cm}^{-2}\ \mu\text{m}^2]$. For values of $a_L \gg 1$ the so called Relativistic Oscillating Mirror (ROM) [3, 4, 15] mechanism is considerably more efficient and the harmonic emission is exclusively due to this process. For intermediate values of $a_L \gtrsim 1$, the two processes can coexist and which one of the two dominates depends sensitively on the shape and gradient of the plasma density profile [12].

The novelty in these investigations is the shortness of the laser pulse in combination with on-target intensities exceeding 10^{18} W cm^{-2}. This constitutes a step toward realization of "intensity gating" for the harmonic emission from solid targets. As in the case of gaseous media, the reduction of the number of cycles under the laser pulse envelope would inevitably lead to fewer attosecond pulses in the generated train [17]. Furthermore, only those cycles within the pulse with enough instantaneous intensity would produce harmonic emission in the necessary spectral range for attosecond time scale emission. In this context, harmonic generation using the plasma medium is more favorable because only one attosecond pulse per laser cycle is produced instead of two as in the case of a gaseous medium. Under these circumstances however, some new effects appear that have not been observed before. For example, it is found that the laser pulses are short enough for effects like random Carrier-Envelope-Phase (CEP) fluctuations to manifest themselves in the observed harmonic spectra. The experimental results are analyzed in terms of a simple model supported by 1D PIC (Particle-In-Cell) simulations. The performed analysis not only explains the main features of the experimental results, but more importantly sheds light on the microscopic electron dynamics responsible for the harmonic emission. The "polarization gating" technique has been also proposed as a method to generate single attosecond pulses [19, 20]. To ascertain the appropriateness of the polarization manipulation to achieve a temporal "gate", a detailed study of the harmonic emission in the CWE regime is undertaken in which (a) the polarization between s- and p-polarized light and (b) the ellipticity of the incident laser pulse was varied. The results are in agreement with PIC simulations and indicate that the first method appears to be quite promising.

2 Experimental investigation of the harmonic emission

The experiments were carried out using the unique few-cycle Light Wave Synthesizer (LWS) [22, 23], which is a Noncolinear Optical Parametric Chirped Pulse Amplification (NOPCPA) system. During these investigations the light source was upgraded from 8 TW (LWS-10) to 16 TW (LWS-20) peak power output. A schematic of the layout of the LWS system is shown in Fig. 1. The Nd:YAG pump laser after frequency doubling delivers 78 ps pulses of 0.5 J in case of LWS-10 and 1 J in case of LWS-20 energy at 532 nm and is optically synchronized to the titanium:sapphire front end (Femtopower, Femtolasers GmbH). The 1 kHz front end produces 25 fs, 800 µJ pulses that are broadened in a hollow-core fiber filled with 2 bar neon to seed the NOPCPA. A grism based stretcher introduces negative group delay dispersion and elongates the pulses to 45 ps that are partially compressed to about 25 ps in an acousto-optical programmable dispersive filter (DAZZLER). The non-collinear optical parametric amplification from 700 nm to 980 nm takes place in two consecutive single-pass Type I BBO stages. In case of the LWS-20 laser system, the pulses are amplified from a few µJ after the DAZZLER up to 1 mJ in the first stage and up to 170 mJ in the second stage. After a bulk compressor including 160 mm SF57 and 100 mm Quartz, an adaptive mirror and a wavefront sensor in a closed loop

Fig. 1 Layout of the Light Wave Synthesizer 20 (LWS-20) laser system [23]

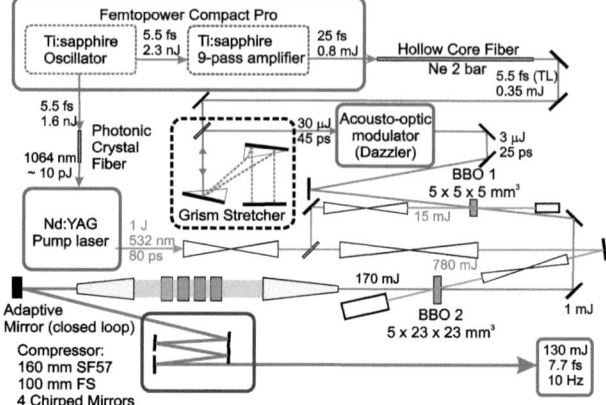

configuration optimize the focusability. The pulses are sent to a vacuum chamber, where four positively chirped mirrors compress them to a typical intensity Full-Width-Half-Maximum (FWHM) duration of 8 fs, which for the 800 nm center wavelength results in sub-3 optical cycles. Accordingly, the E-field of the pulse has a FWHM duration of $\tau_L =$ 4.2 laser cycles. The compressed energy is up to 130 mJ (65 mJ in the case of LWS-10) corresponding to 16 TW (8 TW) power.

Figure 2(a) shows schematically the experimental setup. The laser was focused onto the target at 45° angle of incidence using an F/3, 30° off-axis parabola (effective focal length $f_{eff} = 168$ mm). In case of LWS-20, the average focused intensity inside a focal spot radius containing 86%

of the total energy, was a_L 1.5. The corresponding peak intensity is estimated to be a_L^{peak} 3. The measured laser pulse contrast was 10^{-4} at 3.3 ps and 10^{-8} at 5 ps before the peak [23]. The 120 mm diameter fused silica disc targets were mounted on a rotating mechanism allowing the acquisition of approximately 3500 shots at a repetition rate of up to 10 Hz. The generated harmonic radiation in the specular direction was recorded using a grazing-incidence imaging XUV spectrometer equipped with a CsI coated MCP detector. Except for the measurements where the laser polarization was varied (see Sect. 3), the rest of the data were obtained with p-polarized laser beam. An example of a raw data record is shown in Fig. 2(b). The investigated spectral range was 40–100 nm, which for a laser wavelength of 800 nm corresponds to 8th–20th harmonic orders (H8–H20).

2.1 Harmonic spectrum with 3-cycle laser pulses

Figure 3 shows a set of six single-shot harmonic spectra obtained with our 3-cycle laser pulses in the H11–H16 spectral range. In a more detailed spectral scan a distinct cutoff in the harmonic spectra at H20 was observed. This corresponds to the highest harmonic that the plasma density can sustain confirming that indeed CWE is the dominant harmonic generation process [9]. This is further supported by PIC simulations conducted with parameters close to the ones in the experiment. The measured spectra in Fig. 3 correspond to single laser shots recorded under nominally the same laser conditions. Two distinct characteristics in the spectra are noticeable. The individual harmonics are broadened compared to those from a longer pulse [1, 2, 9, 13] and exhibit from

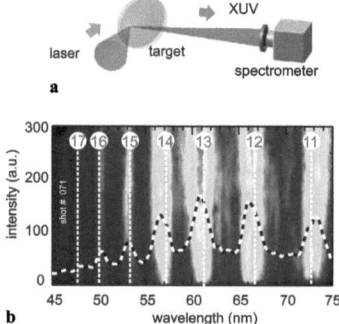

Fig. 2 (a) Schematic of the experimental setup, (b) raw data record obtained with the grazing-incidence imaging XUV spectrometer

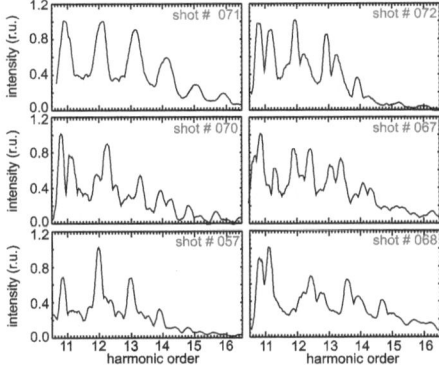

Fig. 3 Single-shot harmonic spectra from fused silica targets. These six records were selected form a set of consecutive shots obtained under nominally same conditions where all the controllable laser parameters were kept fixed. The irreproducible substructure due to CEP variation is clearly visible

shot to shot a strongly irreproducible substructure and a significant spectral shift. Although broader harmonics are expected due to the shortness of the laser pulse, the complex and from shot-to-shot strongly varying spectral structure is a new feature not observed in measurements with longer laser pulses [1, 2]. As discussed in the next Sect. 2.2, the same well behaved spectra exhibiting single harmonic peaks were also observed with the present laser system using positively chirped pulses. This is unequivocal proof that the structure observed with 3-cycle pulses is due to the shortness of the laser pulse.

2.2 Pulse duration variation

The variation of laser pulse duration was accomplished by manipulation of the Group-Delay-Dispersion (GDD) using the DAZZLER incorporated into our laser system (see Fig. 1), i.e., by introducing a chirp in the frequency spectrum of the pulse. This way pulse durations from the nearly Fourier-Transform-Limited (FTL) value of 8 fs to (chirped) pulses of several tens of fs were readily generated.

Figures 4 and 5 depict the main characteristics of the harmonic emission for different pulse durations. They also provide another proof of the fact that the complex structure of the harmonics is due to the shortness of the driving laser pulses. The results indicate that, in contrast to the shortest driving pulse of 8 fs which leads to a complex structure of the harmonics that varies from shot-to-shot (see Fig. 3), the longer pulses obtained by introducing positive chirp values lead to reproducible single peaked harmonics. This is clearly seen in Fig. 4 where a direct comparison of the harmonic spectra for four values of GDD is shown. For nominally FTL pulses (shot # 611), as already discussed, the harmonic spectra appear to be very sensitive to the random variations of CEP when using non-phase-stabilized pulses of only three optical cycles under the envelope (8 fs). For positive values of GDD (shot # 617 and 622) the spectra become progressively narrower and more reproducible. The insusceptibility of the spectrum on the CEP for long, chirped pulses manifest itself in the same way as in the case of experiments on high-harmonic generation on solids and gases with long, FTL pulses. For negative values of GDD (shot # 628) the individual harmonics become even broader and fluctuate even stronger than with FTL pulses. This behavior for different GDD values has been observed before with longer laser pulses [13] and it is attributed to the compensation of the intrinsic harmonic chirp for positive values of GDD. The origin of the intrinsic harmonic chirp is the unequal spacing of the individual pulses in the attosecond pulse train and it is discussed in detail in Sect. 4. In case of uncompensated negative GDD the intrinsic chirp is further increased and the shape of the individual harmonics depends even more sensitively on the CEP variation.

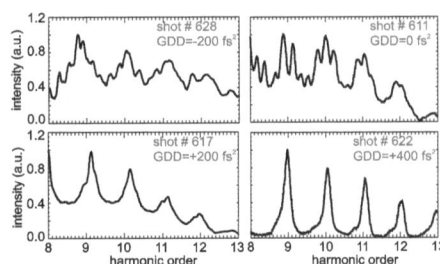

Fig. 4 Harmonic spectra obtained for different values of GDD corresponding to negative chirp (shot 628), no chirp (shot 611) and positive chirp (shots 617 and 622)

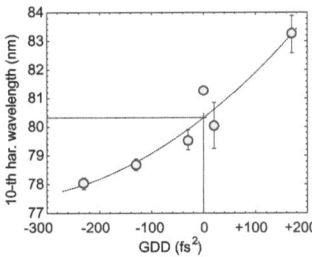

Fig. 5 Dependence of the H10 wavelength on GDD. *The line* is a fit through the experimental points (*circles*)

Another interesting effect observed under variation of the chirp of the laser pulse is the shift of the central wavelength of the individual harmonics. Namely as the GDD value increases the central wavelength also increases. This is depicted in Fig. 5 for the H10. This is attributed to the fact that harmonics are produced predominantly during a particular part of the laser pulse where the intensity is high. As a consequence for a particular value of GDD different parts of the spectrum are responsible for the harmonic generation.

3 Spectral measurements with polarization variation

The dependence of the harmonic emission on the polarization of the incident driving laser pulse is not only interesting from a fundamental point of view but also in connection with the polarization gating technique as a means of generating single attosecond pulses. A number of reports have investigated this subject already theoretically [3, 20, 25, 26] and experimentally [27–31], but no clear picture for the CWE regime has been established so far. In previous reports with long laser pulses or existent prepulses, Rayleigh–Taylor instabilities of the preplasma caused strong surface rippling and therefore polarization scrabbling thus diminishing the effect of changing polarization of the incident laser. However, in our case the shortness and the good contrast of the used laser pulse preserves a clean interaction surface. Since the early theoretical works of Lichters et al. [3] it is well-known that for the polarization dependence of high-harmonics generation certain selection rules apply. In general it was found that the polarization of the generating laser pulse might influence both the intensity and the polarization state of the generated harmonics. In the case of investigations with short-pulse lasers (i.e. shorter than 100 fs) harmonics have only been observed for p-polarized incoming beam. This is expected for the CWE mechanism in which case the existence of an E-field component in the plane of incidence is essential to drag electrons out of the solids (see discussion in Sect. 4). The experiments cited previously aimed primarily at determining the harmonic intensity dependence on the two major polarization directions (s and p) of linearly polarized laser pulses. On the other hand, it is also of interest to investigate experimentally the dependence of the harmonic emission for arbitrarily directed linearly polarized laser light and as a function of the degree of ellipticity. The aim of the present experiment is to obtain a detailed picture of polarization dependence of harmonic generation in the CWE regime. This will provide valuable information regarding the applicability of the polarization gating to generate single attosecond pulses.

The use of the LWS 20 laser system with its polarization insensitive bulk glass and chirped mirror compressor enabled us to change the polarization by inserting small aperture waveplates directly into the unexpanded beam of the laser system where the pulses are still stretched. The polarization state of the laser radiation has been measured at target position, to exclude any influence of the transmission through the compressor and the beamline.

3.1 S–P variation

For measuring the effect of the polarization of the incident laser radiation on the harmonic generation process a commercial broadband, achromatic, zeroth order half-wave plate has been introduced into the laser beam. The plate was rotated in steps of 4 degrees and for every step around 30 spectra in the range of H8 to H13 were recorded with the grazing-incidence spectrometer (see Fig. 2). These spectra have subsequently been integrated over the given harmonic range and averaged. In Fig. 6 the resulting values are shown.

A maximum is observed around fully p-polarized incident laser light. But by rotating the polarization plane to increase the E_s/E_p ratio, the efficiency of the harmonic generation is dropping rapidly to less than 1%, which is around the detection limit of our spectrometer. This behavior is expected, since we are looking at harmonics produced by the CWE mechanism, whereby electrons are accelerated out of the plasma by the component of the incident electric field perpendicular to the target surface. By rotating the half-wave plate in the laser beam, this p-component is decreasing in favor of an increasing s-component while leaving the overall intensity on target constant. This behavior is reproduced by the PIC simulations discussed in detail in Sect. 4.3.

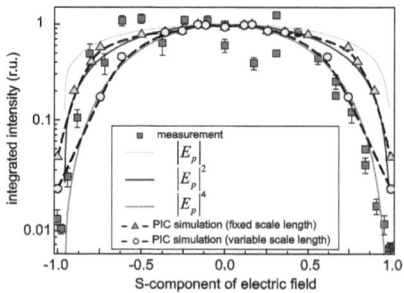

Fig. 6 The effect of the polarization variation from S to P and back to S on the harmonic emission. *The squares* represent the measured values with the error bars being 1 SD over 30 spectra. *The triangles* and *the circles* are the results of PIC simulations with fixed and variable scale length for the preplasma (see discussion in Sect. 4.3). *The curves* indicate the expected dependence $|E_p|^n$ for $n = 1, 2, 4$

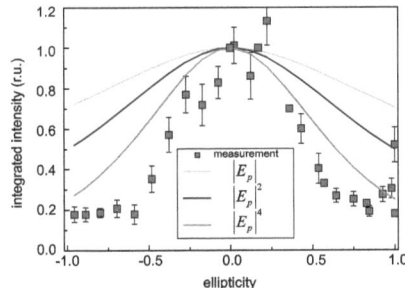

Fig. 7 The effect of the ellipticity variation on the harmonic emission. Points represent the measured values with the error bars being 1 SD over 30 spectra. *The curves* indicate the dependence $|E_p|^n$ for $n = 1, 2, 4$

3.2 Ellipticity variation

Similar to the linear polarization measurement the dependency of the harmonic emission on the incident laser pulse ellipticity has been measured. This measurement was accomplished by replacing the half-wave plate with a zero order quarter wave plate, which also has been rotated in steps of 4 degrees to change the polarization of the transmitted laser beam from circularly polarized on target to linearly p-polarized and back to circular by a rotation of the plate by 90. Again around 30 spectra of the emitted harmonics in the H8–H13 range have been recorded at each position.

In Fig. 7 the integrated and subsequently averaged values of the spectral intensity are shown. As expected, a clear maximum of the harmonic emission is observed for linearly p-polarized light on target. The intensity of the emitted harmonics is dropping rapidly to around 20% of the peak value with increasing ellipticity. Similar to the case of changing from s- to p-polarization, the reduction in efficiency is much faster than expected when considering just the reduction of the component of the driving electric field perpendicular to the target surface (see discussion in Sect. 4.3).

4 Theoretical analysis and discussion

We first investigate some aspects of the Coherent Wake Emission (CWE) mechanism [1, 2, 9, 12–14], which as already discussed, appears to be the dominant harmonic generation process in our experimental studies. In this regime, several features in the emission spectrum can be understood in terms of a *3-step model* that describes the coherent subcycle dynamics of the plasma electrons.

4.1 The 3-step model

According to this model, which is reminiscent of the 3-step model for the harmonic generation in atoms [32], the interaction unfolds as follows: (i) first, the E-field component perpendicular to the target of the obliquely incident, p-polarized laser pulse launches electrons into the vacuum (Brunel electrons [33]), (ii) depending on the E-field phase during the ejection some of these energetic electrons are then hurled back into the plasma during the second half-cycle of the laser period and form bunches that pass through a density ramp, (iii) this gives rise to resonantly driven plasma oscillations at positions x_q within the density gradient where an integer multiple of the driving frequency ω_L coincides with the local plasma frequency ω_p, i.e., where $q\omega_L = \omega_p(x_q)$. At these resonance positions, the excited plasma waves undergo linear mode conversion into EM-waves via inverse resonance absorption [34].

Following the formulation of [33], the electron dynamics can be described in a simplified fashion by the one-dimensional equation of motion along the x-axis (with $x = 0$ on the plasma–vacuum interface). Since we deal with mildly relativistic intensities, i.e., $I_L \lesssim 10^{18}$ W/cm², we write the equation of motion in its relativistically correct form. Each single electron moves in the combined E-field of the laser and electrostatic field due to space charge accumulation:

$$\frac{d\beta}{dt} = 4\pi \sin(\Theta) a_L (1-\beta^2)^{3/2}$$
$$\times \left[E_L(t+t_0) \cos(\omega_L(t+t_0) + \varphi) \right.$$
$$\left. - E_L(t_0) \cos(\omega_L t_0 + \varphi) \right], \quad (1)$$

with x the distance from the interface in wavelengths λ_L, $\beta = dx/dt$. Here $E_L(t) = \exp(-t^2/0.72\tau_L^2)$ is the temporal envelope of the laser E-field, Θ the angle of incidence, and $\omega_L = 2\pi$ the laser frequency. The time t, t_0 and the FWHM intensity pulse duration τ_L are in laser periods T_L and the CE-phase φ is normalized to 2π. The last term in (1) represents the effect of the space charge to the electron motion as formulated in [33]. The instant of release of the electron from the plasma–vacuum interface is denoted as t_0. In the relativistically correct equation of motion we have neglected the magnetic field term to keep the model simple and one-dimensional. As PIC simulation confirm, for mildly relativistic intensities this is justified. A similar analysis but with emphasis on the escaping electrons has been reported in [35]. The numerical solution of (1) for $a_L = 1.5$, $\tau_L = 3$ and $\Theta = 45$ is depicted in Fig. 8(a). The electrons at the vacuum–plasma interface are initially at rest and uniformly distributed, i.e. t_0 is varied in equidistant intervals. After reentry into the plasma region they are assumed to travel at the constant velocity acquired during the excursion

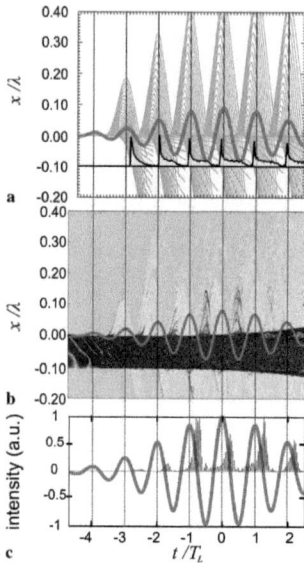

Fig. 8 (a) Numerical solution of (1). The trajectories of the electrons released into the vacuum at uniformly varying values of t_0 and returning to the plasma surface are shown *in green* and the histogram (number of electrons per unit time) at a depth of $x = -0.1$ is plotted *in black line (yellow filled)*. For reference the driving electric field is also plotted (*red line*). (b) The electron density evolution for few cycles at the beginning of the pulse obtained from the 1D-PIC simulation and the E-field of the driving pulse (*red*). The density is color coded with the solid-density target slab appearing as a *dark bar*. (c) The generated as-pulses (*blue*) are positioned at the point in time coinciding with the return of the expelled electrons

in the vacuum. In Fig. 8(a) only those orbits of the electrons returning to the plasma region are plotted. The histogram of the electrons per unit time crossing a line located at $x = -0.1$ behind the interface clearly shows a temporal bunching of the returning electrons within a time window amounting to a fraction of a cycle. The reason for this bunching is merely the electron dynamics for different phases of the sinusoidal E-field of the laser. A careful examination of the orbits in Fig. 8(a) indicates that at the beginning of each cycle (positive-to-negative field crossing) some electrons are pulled out but due to the weak field their excursion is short. As the field continues to grow, the subsequent electrons move further from the surface but due to their higher velocity they return approximately at the same time with the previous ones. This results in a crossover of the orbits inside the plasma and the formation of electron bunches. Since the harmonics are generated via these bunches of energetic electrons (e-bunches), the emission time of the individual

attosecond bursts will follow the occurrence of the electron temporal localization.

We have further substantiated the validity of our simple *CWE 3-step model* by performing simulations using the 1D PIC code PICWIG [20]. The input parameters used were close to those in the experiment, namely $a_L = 1.5$, $\Theta = 45$ and $n_e/n_c = 400$ where n_e is the maximum electron density and n_c is the critical electron density corresponding to the frequency ω_L of the driving laser pulse. The laser spot size is estimated to be $\simeq 7$ μm, which is larger than the length of the 3-cycle laser pulse of $\simeq 2.4$ μm. This indicates that the interaction is nearly one-dimensional and thus the use of 1D PIC code is justified. We chose to conduct the simulations at the average intensity in the interaction area to approximate the behavior of the whole focal region. Furthermore, for the calculation of the harmonic emission a linear density ramp with a scale length of $L = 0.2\lambda_L$ was assumed in front of the plasma slab. This value is based on a simple estimation in which an expansion velocity of 5×10^7 cm/s is assumed and as expansion interval the point of 10^{-4} contrast at 3.3 ps before the peak of the laser pulse, i.e. the point where plasma is generated. The electron density evolution due to a $\tau_L = 3$ (FWHM in intensity) laser pulse of those electrons returning to the plasma for few cycles at the beginning of the pulse is depicted in Fig. 8(b). It exhibits an almost quantitatively similar behavior as the one deduced using the CWE 3-step model (compare Fig. 8(a)). In particular, the time interval during which most of the electrons are pulled into vacuum is preferentially near the first quarter cycle after each positive-to-negative zero crossing of the electric field and their return half a cycle later as predicted by the model also. The attosecond pulse train generated by the coherent superposition of the H10 to H20 harmonics is shown in Fig. 8(c). This confirms the postulation that the harmonic emission contributing to the formation of attosecond pulses occurs at the points in time that coincide with the return of the expelled electrons. From this comparison it becomes apparent that the CWE 3-step model despite its simplicity contains most of the relevant physics.

The change in the oscillation amplitude due to the envelope of the driving pulse gives rise to another effect, which for moderate laser intensities $a_L = 0.2$ and longer pulses has been investigated before using PIC simulations [13]. As is clearly seen in Fig. 8(a), the e-bunches are not equidistant. The distance between them monotonically increases with time while the time interval Δt_n between the nth positive-to-negative zero crossing of the E-field and the resulting e-bunch decreases on the rising edge of the pulse but then increases again. Figures 9(a) and 9(c) show the data points extracted from the corresponding histograms for the unequally spaced e-bunches generated by a 15-cycle and 3-cycle laser pulse. For a Gaussian pulse envelope the data points follow

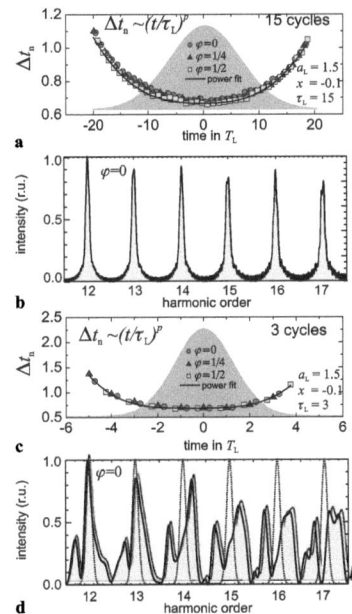

Fig. 9 Time delay between the nth positive-to-negative zero crossing of the E-field and the return of the individual electron bunches calculated for three different CE-phases for (**a**) a 15-cycle and (**c**) a 3-cycle pulse. The intensity of the driving pulses is given for reference. (**b**) and (**d**) show the harmonic spectra corresponding to pulse trains with as-pulses spaced according to the results shown in (**a**) and (**c**) respectively. For comparison, in (**d**) the spectrum for regularly spaced pulses is shown (*dashed line*)

very closely a $\Delta t_n(t,\varphi) = A|(t - (\varphi - \varphi_0))/\tau_L|^p + B$ dependency with A, B, φ_0 and p fitting parameters. The exponent p depends weakly on the pulse duration τ_L and for the two cases considered to a good approximation is constant and equal to 3. This is shown in Figs. 9(a) and 9(c) for the "long" and "short" pulse where Δt_n has been calculated for three CE-phases. It is seen that all values fall on the same curve indicating that it is the pulse envelope that determines Δt_n for each t. Note also that $\varphi_0 \neq 0$ because of the finite time spent by each electron in the vacuum.

4.2 Short-pulse spectra

To assess the influence of the unequal spacing on the spectra expected with laser intensities of $a_L = 1.5$ and for a given pulse duration τ_L we have calculated the spectrum $E_T(\omega)$ of a train of attosecond pulses unequally spaced in time

given by:

$$E_T(\omega) = E_A(\omega) \sum_n E_L(t_n) e^{i\omega t_n}. \quad (2)$$

$E_A(\omega)$ represents the spectrum of a single electron burst of sub-cycle duration that excites plasma oscillations at all frequencies and it is assumed to be constant in the spectral range of interest. According to the CWE 3-step model, for a given φ the attosecond pulse within the nth cycle will occur at time $t_n = \Delta t_n(n,\varphi) + n - (\varphi - \varphi_0)$. Using (2) we have computed the spectrum for $\tau_L = 15$ (Fig. 9(b)) and $\tau_L = 3$ (Fig. 9(d)). For "long" pulses the spectrum exhibits regularly spaced individual harmonics in good agreement with the spectrum from previous experiments [1, 9, 13] and with properly "chirped" pulses (see Fig. 4). When the model is applied to a 3-cycle pulse (see Fig. 9(c)) the effect on the spectrum is striking (Fig. 9(d)). The individual harmonics are not only broadened but also are strongly distorted in shape. It is important to emphasize that despite the distortions the spectrum shown in Fig. 9(d) corresponds to a short train of individual attosecond pulses.

The key to the observed substructure in the harmonic spectrum lies in the intrinsic properties of the CWE generation process, which become perceptible because of the shortness of the driving laser pulse. To investigate the origin of the shot-to-shot variation in the spectrum and assess the impact of various parameters on the spectral structure of the emitted harmonics, we have conducted a parametric study for 3-cycle laser pulses using the model described previously. Using (2) we have calculated the expected spectra for the range $-0.5 \leq \varphi \leq 0.5$ and in the frequency range of the experimentally acquired spectra. Also, for the same laser pulse duration (3-cycles) we have performed PIC simulations using the code LPIC [3] and computed the emitted spectrum similarly for a range of the CEP. The thus obtained spectra for selected values of φ are depicted in Fig. 10 for both cases. We find that for specific values of φ the spectra calculated using the model and PIC simulation reproduce qualitatively remarkably well the features exhibited by the measured ones (see Fig. 3). As seen in Fig. 10 there is no exact correspondence of the spectrum for the same value of φ between model and simulation. This is attributed to the fact that while the PIC simulation computes everything self-consistently the model uses some simplifications. For example, the phase jump between the incident and the reflected wave at the plasma–vacuum interface depends on the local plasma density. This effect is taken into account self-consistently in the PIC simulation. In the model however, a constant phase jump of π is assumed.

According to the model, the time of emergence of the attosecond pulses crucially depends not only on CE-phase φ but also on the overall shape of the Δt_n curve. As expected,

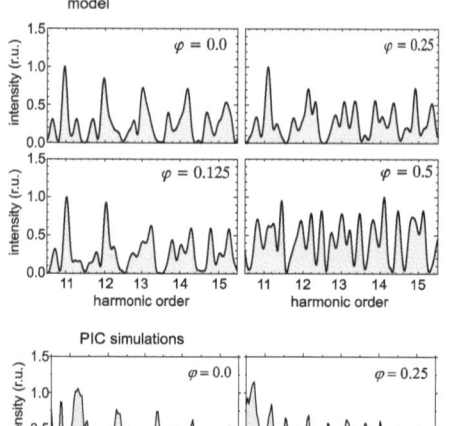

Fig. 10 *Top*: Modeled spectra obtained using (2) in the experimentally recorded range and for four different values of the normalized CE-phase φ. The spacing of the individual as-pulses was inferred from the data shown in Fig. 9. *Bottom*: PIC simulated spectra for the same values of the CE-phase

CEP variations have practically no influence on the harmonic spectrum of the 15-cycle pulse because most of the emitted attosecond pulses emanate in a time interval where the instantaneous intensity variation is negligible and as a result are equidistant. However, this is not the case for the 3-cycle pulses that were used in the experiments. An inspection of the analytical expression for the Δt_n curve shows immediately that its shape would also be affected by inherent pulse duration τ_L fluctuations, which, for our laser system, amount to 3%. Detailed model calculations indicate that this fluctuation level affects the relative amplitude of the peaks within each harmonic, whereas the overall shape (single-vs multi-peaked) is determined predominantly by the value of the CE-phase. A similar analysis shows that intensity (also in the 3% range) and scale-length fluctuations are affecting only the constants A and B and have negligible effect on the calculated spectrum. This suggests that shot-to-shot CEP fluctuations are chiefly responsible for the observed spectral structure.

4.3 Simulations on the effect of the polarization variation in the CWE regime

Previous reports [1, 9] have shown that, for purely p-polarized driver laser, the efficiency of harmonic generation varies with the intensity of the laser as $I_{XUV} \tau_L^{0.4-1} | E_L|^{0.8-2}$. Therefore, for mixed polarization (partially s-polarized and partially p-polarized) one expects that the harmonic intensity would similarly vary as $E_{L,p}^{0.8-2}$ where $E_{L,p}$ is the p-component of the obliquely incident laser field. But from Fig. 6 it is evident, that in our case changing the plane of polarization from purely p to purely s, the efficiency is decreasing considerably faster than $|E_{L,p}|^2$. We have simulated the polarization variation using the PICWIG code and the results are shown in Fig. 6. The input parameters were the same as those given in Sect. 4.1 but the scale length in one case was kept constant at $L = 0.1\lambda_L$ (triangles in Fig. 6) while in the other was reduced progressively from $L = 0.1\lambda_L$ for $E_{L,s} = 0.3$ to $L = 0.04\lambda_L$ for $E_{L,s} = 1.0$ (circles in Fig. 6). As it is discussed later on, this is to simulate the effect of reduced absorption due to increasing s-polarized component in the incident field. As can be seen, the PIC simulations confirm the $|E_{L,p}|^2$ dependence for the case of fixed scale length. On the other hand the reduction of the scale length for increasing values of $E_{L,s}$ yields a dependence much closer to $|E_{L,p}|^4$ and to experimental results.

This result can be understood in terms of the different absorption mechanisms and their dependence on the polarization of the incident light. As has been reported in [36] the absorption of ultrashort laser pulses in overdense plasma is considerably lower in case of s-polarized light. The formation of preplasma and thus the resulting scale length depends on how much energy is absorbed in the time before the arrival of the peak of the pulse. It is reasonable to assume that the amount of absorption decreases as the part of s-polarized component in the incident pulse becomes more dominant. This in turn leads to smaller scale-length values. For reasons discussed by Dromey et al. [37], the consequence of that is the reduction of the harmonic emission due to the CWE mechanism.

In case of ellipticity variation the discrepancy to the simple scaling law shown in Fig. 7 can be attributed to more complex electron dynamics at the vacuum–plasma interface. The field component parallel to the target surface is influencing the trajectories of the extracted electrons. With increasing ellipticity of the incident electric field the electrons emanating from the target surface at different times will feel different parallel field components resulting in trajectories tilted under different angles to the target surface. Electrons reentering the plasma will therefore arrive under various angles which effectively reduces the bunching of these electrons and consequently diminishes plasma wave and with this also harmonic production. Also in this case the absorption and therefore preplasma generation is different for p- and circularly polarized light, which is further affecting harmonic generation. The observation that for circular polarization there is still a 20% harmonics conversion is entirely different from the behavior of gas harmonics. The reason is that for circularly polarized laser light there still exists an electric field component in the plane of incidence, which remains the source of Brunel electrons for the CWE mechanism.

5 Conclusions

In summary, high-harmonic emission from solid targets was studied for the first time using a 3-cycle laser pulse in the multi-TW power scale. New aspects associated with the few-cycle emission have become apparent. We have identified the mechanism responsible for the substructures on the individual harmonics as due predominantly to the sub-cycle dynamics of the electrons responsible for harmonic generation.

The dependence of harmonic conversion on the polarization of the incoming laser beam—in the first approximation—corresponds to the expectations for the CWE mechanism. The high conversion to harmonics for a p-polarized beam drops very fast either by changing the direction of the linearly polarized radiation or by varying the ellipticity. This behavior might be advantageous for a possible approach to the generation of single attosecond pulses using the polarization gating technique. It appears that ellipticity variation is not an appropriate method because the conversion does not vanish even for purely circularly polarized radiation. On the other hand the fast decrease and high contrast when changing the polarization from p to s might open new possibilities for which an s–p–s polarization gating has to be developed.

Although the results obtained with our LWS laser system represent a decisive step forward, to reach the ultimate goal of isolated attosecond pulse generation with reproducible characteristics, further advances in laser technology like multi TW-scale laser systems delivering even shorter pulses with CEP stabilization and excellent characteristics are necessary. Provided that such high-performance laser systems become available, the effort should concentrate in reproducibly generating harmonics via the ROM mechanism, which appears to deliver a harmonic spectrum with superior characteristics, (e.g. higher photon energy, increased efficiency, smaller divergence, better phase locking properties) compared to that from the CWE mechanism.

Acknowledgements This work was funded in part by the DFG projects TR-18 and the MAP excellence cluster, by the LASERLAB-EUROPE, grant agreement # 228334, and by the associations EURATOM—MPI für Plasmaphysik and EURATOM—Hungarian Academy of Sciences. These co-authors gratefully acknowledges financial support as follows: J.M.M. by the Alexander-von-Humboldt Foundation and RFBR grants Nos. 08-02-01245-a and 08-02-01137-a, K.V. by the NKTH-OTKA (#74250) and Janos Bolyai Postdoctoral Fellowships, I.B.F. by the OTKA (#60531), and D.H. by Studienstiftung des deutschen Volkes.

References

1. Y. Nomura, R. Hörlein, P. Tzallas, B. Dromey, S. Rykovanov, Z. Major, J. Osterhoff, S. Karsch, L. Veisz, M. Zepf, D. Charalambidis, F. Krausz, G.D. Tsakiris, Nat. Phys. **5**, 124 (2009)
2. R. Hörlein, Y. Nomura, P. Tzallas, S. Rykovanov, B. Dromey, J. Osterhoff, Z. Major, S. Karsch, L. Veisz, M. Zepf, D. Charalambidis, F. Krausz, G.D. Tsakiris, New J. Phys. **12**, 043020 (2010)
3. R. Lichters, J. Meyer-ter Vehn, A. Pukhov, Phys. Plasmas **3**, 3425 (1996)
4. T. Baeva, S. Gordienko, A. Pukhov, Phys. Rev. E **74**, 046404 (2006)
5. Y.M. Mikhailova, V.T. Platonenko, S.G. Rykovanov, JETP Lett. **81**, 571 (2005)
6. G.D. Tsakiris, K. Eidmann, J. Meyer-ter Vehn, F. Krausz, New J. Phys. **8**, 19 (2006)
7. N.M. Naumova, C.P. Hauri, J.A. Nees, I.V. Sokolov, R. Lopez-Martens, G.A. Mourou, New J. Phys. **10**, 025022 (2008)
8. U. Teubner, P. Gibbon, Rev. Mod. Phys. **81**, 445 (2009)
9. F. Quéré, C. Thaury, P. Monot, S. Dobosz, P. Martin, J.-P. Geindre, P. Audebert, Phys. Rev. Lett. **96**, 125004 (2006)
10. B. Dromey, M. Zepf, A. Gopal, K. Lancaster, M.S. Wei, K. Krushelnick, M. Tatarakis, N. Vakakis, S. Moustaizis, R. Kodama, M. Tampo, C. Stoeckl, R. Clarke, H. Habara, D. Neely, S. Karsch, P. Norreys, Nat. Phys. **2**, 456 (2006)
11. B. Dromey, S. Kar, C. Bellei, D.C. Carroll, R.J. Clarke, J.S. Green, S. Kneip, K. Markey, S.R. Nagel, P.T. Simpson, L. Willingale, P. McKenna, D. Neely, Z. Najmudin, K. Krushelnick, P.A. Norreys, M. Zepf, Phys. Rev. Lett. **99**, 085001 (2007)
12. A. Tarasevitch, K. Lobov, C. Wuensche, D. von der Linde, Phys. Rev. Lett. **98**, 103902 (2007)
13. F. Quéré, C. Thaury, J.-P. Geindre, G. Bonnaud, P. Monot, P. Martin, Phys. Rev. Lett. **100**, 095004 (2008)
14. C. Thaury, H. George, F. Quéré, R. Loch, J.-P. Geindre, P. Monot, P. Martin, Nat. Phys. **4**, 631 (2008)
15. B. Dromey, D. Adams, R. Hörlein, Y. Nomura, S.G. Rykovanov, D.C. Caroll, P.S. Foster, S. Kar, K. Markey, P. McKenna, D. Neely, M. Geissler, G.D. Tsakiris, M. Zepf, Nat. Phys. **5**, 146 (2009)
16. M. Hentschel, R. Kienberger, C. Spielmann, G.A. Reider, N. Milosevic, T. Brabec, P. Corkum, U. Heinzmann, M. Drescher, F. Krausz, Nature **414**, 509 (2001)
17. F. Krausz, M. Ivanov, Rev. Mod. Phys. **81**, 163 (2009)
18. P. Tzallas, E. Skantzakis, E.P. Benis, G.D. Tsakiris, D. Charalambidis, Nat. Phys. **3**, 846 (2007)
19. T. Baeva, S. Gordienko, A. Pukhov, Phys. Rev. E **74**, 065401 (2006)
20. S. Rykovanov, M. Geissler, J. Meyer-ter Vehn, G.D. Tsakiris, New J. Phys. **10**, 025025 (2008)
21. A. Baltuska, M. Uiberacker, E. Goulielmakis, R. Kienberger, V.S. Yakovlev, T. Udem, T.W. Hänsch, F. Krausz, IEEE J. Sel. Top. Quantum Electron. **9**, 972 (2003)
22. F. Tavella, Y. Nomura, L. Veisz, V. Pervak, A. Marcinkevicius, F. Krausz, Opt. Lett. **32**, 2227 (2007)
23. D. Herrmann, L. Veisz, R. Tautz, F. Tavella, K. Schmid, V. Pervak, F. Krausz, Opt. Lett. **34**, 2459 (2009)

24. C. Thaury, F. Quéré, J.-P. Geindre, A. Levy, T. Ceccotti, P. Monot, M. Bougeard, F. Réau, P. d'Oliveira, P. Audebert, R. Marjoribanks, P. Martin, Nat. Phys. **3**, 424 (2007)
25. P. Gibbon, Phys. Rev. Lett. **76**, 50 (1996)
26. K. Gál, S. Varró, Opt. Commun. **198**, 419 (2001)
27. L.A. Gizzi, D. Giulietti, A. Giulietti, P. Audebert, S. Bastiani, J.P. Geindre, A. Mysyrowicz, Phys. Rev. Lett. **76**, 2278 (1996)
28. E. Rácz, I.B. Földes, G. Kocsis, G. Veres, K. Eidmann, S. Szatmári, Appl. Phys. B, Lasers Opt. **82**, 13 (2006)
29. P.A. Norreys, M. Zepf, S. Moustaizis, A.P. Fews, J. Zhang, P. Lee, M. Bakarezos, C.N. Danson, A. Dyson, P. Gibbon, P. Loukakos, D. Neely, F.N. Walsh, J.S. Wark, A.E. Dangor, Phys. Rev. Lett. **76**, 1832 (1996)
30. G. Veres, J.S. Bakos, I.B. Földes, K. Gál, Z. Juhász, G. Kocsis, S. Szatmári, Europhys. Lett. **48**, 390 (1999)
31. R.A. Ganeev, A. Ishizawa, T. Kanai, T. Ozaki, H. Kuroda, Opt. Commun. **227**, 175 (2003)
32. P.B. Corkum, Phys. Rev. Lett. **71**, 1994 (1993)
33. F. Brunel, Phys. Rev. Lett. **59**, 52 (1987)
34. Z.-M. Sheng, K. Mima, J. Zhang, H. Sanuki, Phys. Rev. Lett. **94**, 095003 (2005)
35. F. Brandl, B. Hidding, J. Osterholz, D. Hemmers, A. Karmakar, A. Pukhov, G. Pretzler, Phys. Rev. Lett. **102**, 195001 (2009)
36. M. Cerchez, R. Jung, J. Osterholz, T. Toncian, O. Willi, Phys. Rev. Lett. **100**, 245001 (2008)
37. B. Dromey, S.G. Rykovanov, D. Adams, R. Hörlein, Y. Nomura, D.C. Caroll, P.S. Foster, S. Kar, K. Markey, P. McKenna, D. Neely, M. Geissler, G.D. Tsakiris, M. Zepf, Phys. Rev. Lett. **102**, 225002 (2009)

i want morebooks!

Buy your books fast and straightforward online - at one of world's fastest growing online book stores! Environmentally sound due to Print-on-Demand technologies.

Buy your books online at
www.get-morebooks.com

Kaufen Sie Ihre Bücher schnell und unkompliziert online – auf einer der am schnellsten wachsenden Buchhandelsplattformen weltweit! Dank Print-On-Demand umwelt- und ressourcenschonend produziert.

Bücher schneller online kaufen
www.morebooks.de

VDM Verlagsservicegesellschaft mbH
Heinrich-Böcking-Str. 6-8   Telefon: +49 681 3720 174   info@vdm-vsg.de
D - 66121 Saarbrücken   Telefax: +49 681 3720 1749   www.vdm-vsg.de

Printed by Books on Demand GmbH, Norderstedt / Germany